Memoirs Of Hans Hendrik, The Arctic Traveller, Serving under Kane, Hayes, Hall and Nares, 1853-1876

Written by himself.

Translated from the Eskimo language by Dr. **HENRY RINK** **[Dr. Hinrich Johannes Rink]**, Director of the Royal Greenland Board of Trade, Author of "Tales and Traditions of the Eskimo," "Danish Greenland," &c.

Edited by Prof. Dr. **GEORGE STEPHENS**, F.S.A., Lond., Edin., and Stockholm, &c., &c.

Original publication: London: Trübner & Co., Ludgate Hill. 1878.

ISBN: 9798656132589

Cover: Hans Hendrik 'the Esquimaux' with his daughter/wife[?] and son on the upper deck of 'Discovery' (1873).

Author portrait: Ida Falander: Porträt des grönländischen Inuk Hans Hendrik, Teilnehmer an fünf Polarexpeditionen im Jahr 1883 nach einer Fotografie

This work is in the public domain.

Editor's Note: The original language has been preserved, including outdated terms.

Table of Contents

INTRODUCTION..5

THE TALE OF MY TRAVELS TO THE HIGH NORTH.......28

MY SECOND NORTHERN JOURNEY..................................42

MY THIRD JOURNEY TO THE NORTH..............................52

MY FOURTH VOYAGE TO THE NORTH, WHEN I WAS ENGAGED BY THE TULUKS..85

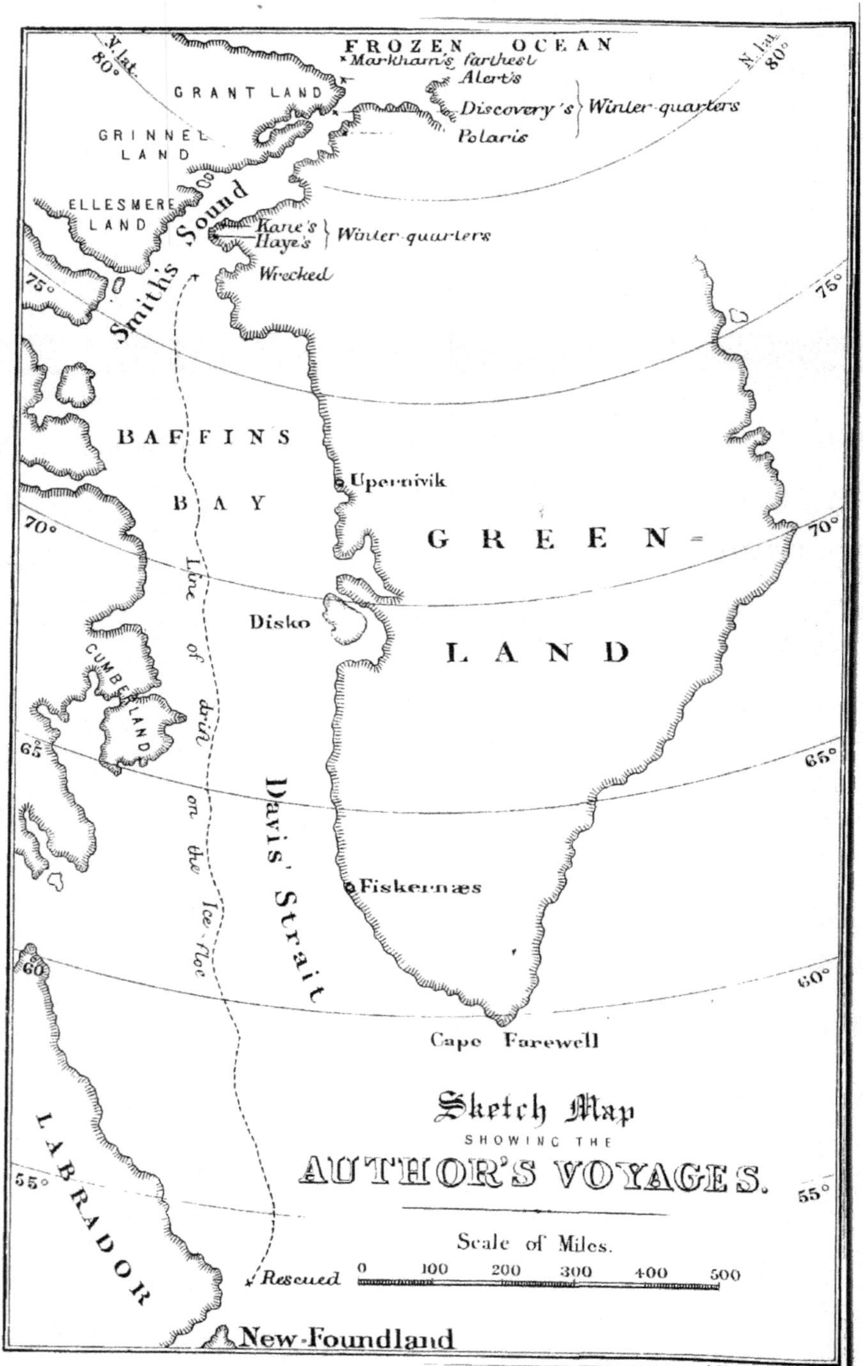

FROZEN OCEAN
Markham's farthest
Alert's
Discovery's
Polaris
Winter-quarters
N. lat. 80°
N. lat. 80°
GRANT LAND
GRINNEL LAND
ELLESMERE LAND
Smith's Sound
Kane's
Haye's
Winter-quarters
Wrecked
75°
75°
BAFFINS BAY
Upernivik
70°
70°
GREEN-LAND
Disko
CUMBERLAND
Line of drift on the Ice-floe
Davis' Strait
65°
65°
Fiskernæs
60°
60°
Cape Farewell
Sketch Map
SHOWING THE
AUTHOR'S VOYAGES.
55°
55°
LABRADOR
Scale of Miles.
0 100 200 300 400 500
Rescued
New-Foundland
PETTITT'S Co. LITH. LONDON

INTRODUCTION

A couple of months ago I received from my friend, Herr Krarup Smith, who resides in Disko Island, a narrative written last winter in the Eskimo language by a native who had shared in several Arctic expeditions. Herr Smith, who is Inspector of the Northern Danish Settlements in Greenland, supposed that parts of the MS. might be fit for publication in some journal. He therefore suggested that I should make such extracts as might suit this purpose. But I had hardly run over the pages before I had made up my mind to publish it entire, just as it was. What I have struck out is not worth mentioning. My reason was, that I had never read any adventures in the far North so curious relatively to their shortness. On translating specimens to others I was corroborated in my opinion, and especially we agreed that, besides setting forth striking vicissitudes, every line helps to describe the inhabitants of the Arctic regions, by reflecting their ideas and their mental development in the person of our author.

I was led to undertake the difficult task of translating the MS. into English, not only by the desire to render it accessible to the widest circle possible, but also from a special regard to our author's fellow-travellers in England and America. My doubt whether I should be able to render the sketch tolerably well in English was overcome by my friend, Professor G. Stephens, offering to give it a final revise.

This peculiar record requires some explanation, both as to the author himself, and as to the renowned travellers whom he accompanied.

Birthplace and Nationality of the Author

In Southern Greenland, on the border of Davis Strait, is the small trading establishment, Fiskernæs. Its latitude is not more northerly than middle Norway, but its climate is more severe than the northernmost coasts of Norway and Iceland. It has been proved by experience that, nowadays, only the present natives are able to live even in the most favourable tracts of Greenland, without being supplied with their chief necessaries from more genial countries. Our author belongs to that remarkable Eskimo race which is spread from Greenland to Behrings Straits, and is able to procure a comfortable existence in countries where men of our race only have been able to stay for a couple of years by the help of the numerous and expensive resources of modern civilisation.

It is well known that the capability of the Eskimo to brave their climate depends on their ingenuity in catching and making use of the seal. When they find no better materials, they build a comfortable house merely out of snow, both light and heat it with their seal-oil lamps, manufacture excellent garments out of sealskins, and have the most suitable food for a cold climate in the flesh and blubber of the same animals.

Our author affords a striking example of the independence of his nation, of the climate within their vast territories, as well as of aid from foreign nations. When a young man he was suddenly removed from his home to a country about 1000 miles nearer the North Pole, and found himself so attracted by its amenities, that he did not hesitate to settle down there. Furthermore, in his birthplace his countrymen are accustomed to have in their immediate vicinity a shop where they may barter their produce for all sorts of European articles. In Greenland we divide this merchandise into what is necessary, useful, or a luxury. But he proved the whole to be nearly superfluous, for he settled amongst a tribe not only in a state of perfect seclusion, but which had scarcely seen a white man before.

Fiskernæs comprises the trading post of the same name and the Moravian missionary station, Lichtenfels, upon an island, some 3 miles from each other, and numbering both together

240 inhabitants. For more than 100 years Lichtenfels has been the residence of from two to four missionaries, who are recruited from Germany. But the natives here are very poor, and the community has decreased nearly one-half in the last thirty years. The most obvious feature in their impoverishment is their want of boats for their travelling life in summer. This roaming is necessary not only with regard to their hunting and fishing, but also for their health. It counteracts the deadening influence of the climate, and the isolated situation of the dwelling-places. The natives of Lichtenfels only exceptionally have been farther than 20 miles from their home, and many, perhaps, never leave it. I note this expressly to throw light on the condition of our author when he was engaged by the foreign travellers.

Seal-hunting by kayak is still continued by the Greenlanders in the same way as by their ancestors a thousand years ago. The strangers who settled in their land have not taught them the least improvement as regards this chief means of subsistence. For this reason, and as they have kept their language unaltered, the Greenlanders maintain a certain independence, notwithstanding the general supremacy of foreigners. They know that they must wholly rely upon themselves; and their peculiar life under numerous hardships and dangers develops from early youth a faculty of self-help not so often found in civilised societies, where division of labour prevails. For the same reason, poverty has a less depressing influence than elsewhere. The hunter will always keep up a certain degree of mental vivacity contrasting with his impoverished state. The dangers besetting kayak-hunting are especially bracing.

The Greenlanders have also taken well to school instruction, and skill in reading and writing is as common amongst them as in any other country.

The same contempt with which white men look down upon people of other races has amply manifested itself in his intercourse with the Eskimo. It has been asserted a hundred years ago, that in Greenland the worst Dane was considered better than the best Greenlander; and this may be so even occasionally now. If a man brought up as a native seal-hunter takes service with foreigners, many of whom consider him an inferior being, and who can only speak with him imperfectly by interpreters of the superior race, he at times must feel himself misunderstood and wronged. A native like Hans, who was taken from his quiet and solitary homestead and had to live with so many strangers, could not help at times being placed in this condition. What he says on such occasions will therefore be found a natural part of the picture he gives. However, thoroughly to understand the strange suspicions exhibited in some parts of his statement, we must consider the traditions still living amongst the Greenlanders about atrocities formerly committed in their country by foreigners, as well as their indistinct ideas of the wars and military discipline of the white men.

But we see that the instances of feeling himself aggrieved were exceptional. That mutual satisfaction was the rule, is also evident from his taking employ so often. As regards his superiors, I shall only remark that the first of them, Kane, praises him as a very useful and active fellow, on whose energies as a hunter the supplies of the travellers often depended; and the last of them, Sir George Nares, says in his official report: "All speak in the highest terms of Hans the Eskimo, who was untiring in his exertions with the dog-sledge and in procuring game."

*The American Expedition under Kane
1853-1855.*

The small sketch map which I have added to this book gives an idea of our author's Arctic travelling routes. It will be seen that he shared in all the four renowned expeditions which have thrown light upon the narrow branch of the sea that divides the Greenland from the American Arctic islands. As far as I know he is the only man who did so. These voyages were undertaken with very different resources, but all of them exhibit examples of skill, courage, and perseverance rivalling the most daring enterprises in other parts of the Arctic regions. The final result with regard to the North Pole, that they only explored the whole of Smith's Sound, proves the enormous difficulties they had to surmount in forcing their way step by step through this passage. The chief aim, however, has now been attained. The end of this mysterious thoroughfare has been reached, and it was found to lead to an ocean, without any land visible, and covered with ice apparently moving only at intervals of many years and without the least probability of being navigable, while on the other hand the extraordinary ruggedness of the ice defies the sledge.

The pioneer was Elisha Kent Kane, who had been appointed Surgeon to the Grinnel Expedition in search of John Franklin in the years 1850 to 1851. As the leader of a new expedition he left New York on the 30th of May 1853, in the brig 'Advance.' He touched at Fiskernæs, where he engaged Hans; and at Upernivik, where he obtained another worthy helper, a Dane named Carl Petersen, who became famous afterwards as interpreter and assistant to Arctic navigators.

The other members of the company were—

>Henry Brooks, *First Officer.*
>Isaac J. Hayes, M.D., *Surgeon.*
>August Sonntag, *Astronomer.*
>John Wall Wilson.
>James McGary.
>George Rilay.
>William Morton.

Christian Olsen.
Henry Goodfellow.
Amos Bonsall.
George Stephenson.
George Whipple.
William Godfrey.
John Blake.
Jefferson Baker.
Peter Schubert.
Thomas Hickey.

Kane took his winter-quarters in Rensselaer Harbour, lat. 78° 30'. This place being 15° north of Fiskernæs, where during the shortest day the sun appears about two hours above the horizon, it is not wonderful that Hans was surprised by a perpetual night of about four months' duration. His ideas of the northernmost regions must have been derived from the tales and traditions of the Greenlanders; and these, as far as I remember, seldom mention the darkness in winter, but chiefly dwell on the severity of the weather. However, I am inclined to believe that his terror was partly connected with his ideas, about the destruction of the world, which may have been rooted in his imagination at an early age.

From his two years' winter-quarters on the eastern border of the sound, Kane, by help of dog and drag sledges, undertook surveys in different directions, partly across to the American side, but mainly along the coast, pursuing it northward, to find, if possible, the northern end of Greenland. Of these sledge expeditions, that of Morton and Hans reached the highest latitude—80° 40'—and discovered the open water long said to be connected with an open sea around the North Pole.

In his travelling account, Dr. Kane goes beyond what he promises in his preface—a simple narrative of the adventures of his party. He has set forth theories which afterwards proved unfounded. This, of course, has injured his work, but not cast into the shade the merits of this gallant and talented man, who brought us the first intelligence from those inhospitable regions. His book gave numerous interesting and remarkable facts concerning northernmost Greenland, obtained by immense labour and rare efforts. Especially is his information valuable on the mode of life of the small isolated population of all mankind nearest to either Pole, braving the climate from the abundance of walruses, bears, and other animals, first described by Kane.

It is well known that Kane, in 1855, abandoned his vessel, and made his escape in boats to the most northerly Danish trading station, Upernivik. Only Hans remained, settling amongst the natives of Smith's Sound.

*The American Expedition under Hayes
1860—1861.*

I. J. Hayes, who acted as Surgeon of the expedition commanded by Dr. Kane, after his return, drew up a plan for a new exploring voyage towards the North Pole, *viâ* Smith's Sound. With this object he applied to the scientific public for assistance; first the American Geographical and Statistical Society, before which he read a paper in December 1857, setting forth his scheme, and the means proposed for its accomplishment. Notwithstanding his indefatigable zeal, he did not succeed in fitting out an expedition by aid of private subscription before 1860. On July 7th he left Boston, in the schooner 'United States,' accompanied by the Astronomer, Sonntag, who had been with Kane.

The whole list of the company was—

> I. J. Hayes, *Commander.*
> August Sonntag, *Astronomer and second in command.*
> S. J. McCormick, *Sailing Master.*
> Henry W. Dodge, *Mate.*
> Henry G. Radcliffe, *Assistant Astronomer.*

George F. Knorr, *Commander's Secretary.*
Collin C. Starr, *Master's Mate.*
Gibson Caruthers, *Boatswain and Carpenter.*

Volunteers.
Francis L. Harris and Harvey Heywood.

Seamen.
Thomas Barnum.
John McDonald.
Charles McCormick.
William Miller.
John Williams.

The ship touched at Upernivik and at the outpost Tasiusak, northward of this settlement, where the party was increased to twenty-one souls, namely, by—

Danes. Peter Jensen, *Interpreter and Dog-manager.*
Carl Emil Olsvig, *Sailor.*
Carl Christian Petersen, *Sailor and Carpenter.*

Eskimo Hunters and Dog-drivers.
Peter, Marcus, Jacob.

At the end of August they touched at Cape York and picked up Hans and his wife and child.

Hayes did not bring his ship quite as far as Kane, the hindrances from the drift-ice proving still greater this year. But the difference was only small; and, on the other hand, it must be remembered that Hayes brought his ship safe home. He took winter-quarters in Port Foulke, 40 miles S.W. of Rensselaer Harbour. On November 19th, the native Peter, or "Pele," disappeared; and on December 21st, just in the middle of the winter night, two months before the sun should re-appear, Hans and Sonntag started on their fatal journey, from which the latter was never to return. As alluded to by Hans, it was suggested by some on board that he had caused Peter to run away, and given false report as to Dr. Sonntag's death, which he had not taken pains enough to prevent. I only mention this as a curiosity, in connection with what I have stated above on the relations between the Eskimo and the foreigners in general.

On April 3rd, Hayes set out with a party consisting of twelve men, having one dog-sledge and one hand-drawn sledge, to cross Smith's Sound, and proceed as far as possible to the north on the other side. The difficulties were of the most disheartening kind. The whole sound was filled with drift-ice of the most massive description, crushed, piled up, and now frozen together, and having the spaces and crevasses between the hummocks more or less filled with drifted snow. It was like an accumulation of rocks closely packed and heaped up over a waste plain in great clusters and endless ridges, leaving scarcely a foot of level surface. The travellers had to pick their way as best they could amongst the inequalities, crawling over the walls of screwed ice and sometimes even climbing bergs more than 50 feet high. The obstacles encountered may be judged of by the fact that the journey of 80 miles across the sound took thirty-one days, notwithstanding the utmost exertions of strong and persevering men. The drag-sledge was left behind, and at last Hayes retained only one companion, George Knorr, with whom he reached his furthest point on the 18th of May, supposed to be a half degree farther north than the latitude reached by Hans and Morton on the opposite side in 1854.

July 14th, 1861, the ship was released and left Port Foulke. August 15 th, it anchored in the harbour of Upernivik.

The American 'Polaris' Expedition under Hall 1871—1873.

Charles Francis Hall, a native of Cincinnati, had made two previous voyages to the Arctic regions, and by a long residence amongst the Eskimo thoroughly acclimatized himself, and acquired a complete knowledge of Eskimo life. He was neither seaman nor possessed of scientific acquirements, but an enthusiastic leader who had wholly devoted himself to Arctic discovery. For the expedition now in question he had a whaling Captain, Buddington, as his sailingmaster. Dr. Bessels who had studied in Heidelberg, and had already made one trip to the Arctic regions, was the leader of the scientific explorations. The 'Polaris,' to whose equipment Congress had contributed 30,000 dollars, left New York June 29th 1871. The last letter from Hall was dated Tasiusak in lat. 73° 21', August 24th. The passage north proved so exceptionally favourable this year, that from about the latter place the 'Polaris' reached her furthest northern point in Smith's Sound, lat. 82° 12', in five days.

Here they were stopped, and though Hall was very reluctant to turn back, they followed the advice of the sailing-master and took winter quarters in Polaris Bay in lat. 81° 27'. Eighteen miles north of this place the ice appeared heavier than anywhere further south, but all visible from the harbour seemed only of one year's growth. From their furthest point they believe they saw land in a northerly direction, stretching as far as 84°. Hall died of apoplexy in November 187 1. His attack was caused by want of caution in exposing himself to a sudden change of temperature on a sledge journey. The relation which Hans gives of his wife's sickness and vision is very characteristic with regard to the belief of the natives in the so called "Kivigtoks." By this name are denoted persons who flee mankind, taking refuge in the desolate interior of Greenland. By living alone with nature they are supposed to obtain supernatural faculties. Their senses are sharpened so as to enable them to perceive what passes many miles off, they understand the speech of animals, and move with the quickness of a bird's flight. The reasons which lead men to become Kivigtoks are supposed to be unjust treatment by their house-mates or relatives, who in this case always were considered in danger of vengeance from the hand of the fugitive. People who are in a state of madness or, as Hans's wife in this instance, of delirium, are considered especially able to communicate with supernatural beings and to see absent and concealed things. The disappearance to which her vision refers, must have taken place at Pröven, the station south of Upernivik, where they lived before they embarked in the 'Polaris.' It also seems a supernatural feature that she had forgotten her meeting with the Kivigtok, until she grew delirious.

The 'Polaris' left her winter-quarters in the summer of 1872, was beset, and drifted down Smith's Sound. It was on the 15th of October that the movement of the frozen mass in a heavy gale caused the ship's crew to land boats and provisions on the ice, to be prepared for the worst. In the following night the accident occurred which separated the ship's company.

The following persons were left upon the ice:—

> Tyson, *Second Captain.*
> F. Meyer, *Assistant to Dr. Bessels.*
> John Herron.
> J. W. Kruger.
> W. Jackson.
> W. Lindermann.
> Peter Johnson.
> F. Anthing.
> G. W. Linguist.
> Fred. Jamka.
> Joe, Eskimo from the West Coast, with his wife, Hanna, and one foster-daughter [?].
> Hans and his wife, with (as far as I can make out) the following four children:—Augustina, Tobias, Sophia, and Charlie Polaris (aged two months).

The following persons were drifted off with the ship:—

> Captain Buddington.
> Dr. Bessels.
> Capt. Chester, *First Mate.*
> Mr. Morton, *Second Mate.*
> Mr. Schumann, *Engineer.*
> Mr. Odell, *Second Engineer.*
> Nathan Coffin, *Carpenter.*
> Noah Hayes.
> Herman Siemens.
> Henry Hobby.
> W. F. Campbell.
> Mr. R. Bryant, *Assistant Astronomer.*
> Jos. Maneh.
> John Booth.

The 'Polaris' was run ashore in the entrance of Smith's Sound, where the party of fourteen passed the winter. They built two boats out of the wreck, left their winter-quarters on the 4th of June, and were picked up on the 23d of June by the whaling vessel 'Ravenscraig.'

The party left upon the ice were rescued on the 30th of April 1873 by the 'Tigress,' in lat. 53° 35', not far from Newfoundland. Their ice-floe, which at first had a few miles in circumference, was successively reduced to a diameter of a hundred yards. Upon this they passed six months, nearly half of which must have been perpetual night, and they travelled at the same time about 1500 miles over the ocean. As far as I can make out, they should have perished, if they had got no seals that could supply them with oil for their lamps; but I have searched in vain for information, whether the two Eskimo men were the only members of the party who caught them; and, if not, how many were killed by the other ten.

The English Expedition tinder Captain Nares
1875—1876.

The ships 'Alert,' commanded by Captain Nares, leader of the expedition, and 'Discovery,' commanded by Captain Stephenson, left England on the 29th of May 1875. Smith's Sound was not found more free than usual, and, considering the sacrifice of time and labour hitherto on the exploration of this inlet, we shall be able to appreciate the masterly manœuvre by which the vessels were conducted through it. The constant drift of the heavy floes, which closed on each other with great force, required constant watchfulness, lest the ships, caught between them, should be instantaneously crushed. Keeping to the west side of the sound, they had to be secured in the best way possible when the road appeared quite blocked up, while every favourable moment for proceeding had to be used, and a very tortuous course to be followed.

In about the extreme latitude that had been reached by the 'Polaris,' the 'Discovery,' on board of which was our author, took, her winter-quarters in a convenient harbour. The 'Alert' then continued coasting northward, but now met with drift of extraordinary dimensions, exhibiting a thickness of from 40 to 80 feet. They succeeded, however, in reaching the very end of the sound, and what lay beyond it now turned out to be a frozen ocean, without any land to be sighted to the north. It was from this icy floor that the heavy pack was detached which had been met with during the last part of the voyage. Off the open coast facing this frozen sea the ship anchored, sheltered by a row of ice-blocks. These, from their enormous thickness, had grounded, in this way forming her winter harbour in lat. 82° 30'.

Sledge parties were sent out from both ships in autumn, but especially in spring, to explore in three directions—eastwards the coast of Greenland; westwards, the north coast of the

American Arctic islands; and northwards, the frozen sea, trying to penetrate it as far as possible.

To the east, Lieutenants Beaumont and Rawson reached a cape which in all probability is the northernmost point of Greenland, or, at least, near to it. As these explorations of northernmost Greenland, opposite the winter-quarters of the two ships, were those in which Hans shared, I insert the following abstract of Captain Nares' report concerning them: —

"On April 18, Lieut. Rawson and Mr Egerton returned, having succeeded in crossing the channel without finding more than the usual difficulties amongst the heavy hummocks, which they had now become so accustomed to. They had landed on the Greenland coast north of the position marked as Repulse Harbour, which proves to be only a slight indentation in the coast line, having a fresh-water lake inshore of it, which from an inland view might readily be mistaken for a harbour.

"On April 20, Lieut. Beaumont, accompanied by Lieut. Rawson and Dr. Coppinger, started for his Greenland exploration, the few days' rest having materially benefited his men, who may be said to have started from the 'Discovery' unexperienced in Arctic sledging, that ship having had no autumn travelling in consequence of the ice remaining in motion until a very late period of the season.

* * * * * * * * *

"On May 9, by the return of Lieut. May and Mr. Egerton from Greenland, whither they had carried supplies and succeeded in discovering a practicable overland route immediately east of Cape Brevort fit for the use of the returning sledges should the ice break up, I received news of Lieut. Beaumont's party up to May 4, when he was within two miles of Cape Stanton. From their place of crossing the Straits they found that the coast line for nearly the entire distance to Cape Stanton was formed either by precipitous cliffs or very steep snow slopes, the bases of which receive the direct and unchecked pressure of the northern pack as it drifts from the north-westward and strikes against that part of the coast nearly at right angles. The floe-bergs, at their maximum sizes, were pressed high up one over the other against the steep shore; the chaos outside was something indescribable, and the travelling the worst that can possibly be imagined, seven days being occupied in moving forward only twenty miles. Being quite uncertain when such a road might become impassable by the ice breaking up in May as it did in 1872, a depôt of provisions, sufficient for a return journey by land, was wisely left, but Lieut. Beaumont's journey was thus shortened considerably.

* * * * * * * * *

"On June 1, Mr. Crawford Conybeare arrived with news from the 'Discovery' up to May 22. Lieut. Archer had completed his examination of the opening in the land west of Lady Franklin Sound, proving it to be a deep fiord terminating in mountainous land, with glacier-covered valleys in the interior.

"Lieut. Reginald B. Fulford, with the men returned from Lieut. Archer's party, then transported two boats across Hall's Basin to assist Lieut. Beaumont in his return later in the season. Capt. Stephenson accompanied by Mr. Henry C. Hart, naturalist, overtook this party on the 12th at Polaris Bay. On the following day, the American flag being hoisted, a brass tablet prepared in England was erected at the foot of Capt. Hall's grave with due solemnity. It bore the following inscription:—

'Sacred
to the Memory of
Captain C. F. Hall,
of the U.S. Ship 'Polaris,'
who sacrificed his Life
in the advancement of Science,

on the 8th November, 1871.
'This tablet has been erected by the British Polar Expedition of 1875, who, following in his footsteps, have profited by his experience.'

"Dr. Coppinger, when returning from assisting Lieut. Beaumont, had visited Capt. Hall's Cairn at Cape Brevort, and the boat depot in Newman's Bay, and conveyed the few articles of any value to the 'Discovery.' The boat itself, with the exception of one hole, easily repairable, was in a serviceable condition. Capt. Stephenson returned to the 'Discovery' on May 18, leaving Lieut. Fulford and Dr. Coppinger on the Greenland shore to explore Petermann Fiord. Mr. Crawford Conybeare having reported that the travelling along shore in Robeson Channel was fast becoming impracticable in consequence of the ice being in motion near the shore, his party were kept on board the 'Alert.'

* * * * * * * * *

"On August 6, while the 'Alert' was imprisoned by the ice twenty miles north of Discovery Harbour, during her passage down Robeson Channel, Lieut. Rawson and two men arrived with letters from Capt. Stephenson containing the distressing intelligence that scurvy had attacked the Greenland Division of sledges with as much severity as it had visited the travellers from the 'Alert,' and that Lieut. Beaumont was then at Polaris Bay recruiting his men. I must refer you, Sir, to Capt. Stephenson's letters and to Lieut. Lewis A. Beaumont's report for a full detail of the proceeding of this party, but I may here mention the chief points. I have already reported their movements up to May 5, when Dr. Coppinger left them; Lieut. Beaumont with two sledge crews journeying to the north-eastward along the north coast of Greenland, all apparently in good health. A very few days after, James J. Hand, A.B., who had passed the winter on board of the 'Alert,' showed symptoms of scurvy. As soon as the nature of the disease was decided, Lieut. Beaumont determined to send Lieut. Rawson with three men and the invalid back to Polaris Bay, and to continue the exploration with reduced numbers. Lieut. Wyatt Rawson parted company on his return on May 11; but owing to two more of his crew breaking down, leaving only himself and one man strong enough to drag the sledge on which lay the principal sufferer, and to look after the other two, he only succeeded in reaching the depôt on June 3, James J. Hand unhappily dying from the extreme fatigue a few hours after the arrival of the party at Polaris Bay. Out of the other men forming the sledge crew, who had all passed the winter on board the 'Alert,' only one of them—Elijah Rayner, Gunner, R.M.A.—escaped the insidious disease; George Bryant, 1st class petty officer and captain of the sledge, and Michael Regan, A.B., were both attacked, the former, although in a very bad state, manfully refused to the last to be carried on the sledge, knowing that his extra weight would endanger the lives of all.

"I cannot praise Lieut. Rawson's conduct on this occasion too highly; it is entirely due to his genial but firm command of his party, inspiriting as he did his crippled band, who relied with the utmost confidence on him, that they succeeded in reaching the depôt. His return being totally unexpected, no relief was thought of, nor, indeed, were there any men to send. On June 7 Lieut. Fulford and Dr. Coppinger, with Hans and the dog-sledge, returned to Polaris Bay depôt from the exploration of Petermann Fiord; and, with the help of some fresh seal meat and the professional skill and care of Dr. Coppinger, the malady was checked and the sick men gradually regained strength.

"Lieut. Beaumont, continuing his journey on May 21, succeeded in reaching lat. 82° 18′ N., long. 50° 40′ W., discovered land, apparently an island, but, owing to the nature of the ice, probably a continuation of the Greenland coast, extending to lat. 82° 54′ N., long. 48° 33′ W. By this time two more of the crew showed symptoms of scurvy, and soon after the return journey was commenced the whole party were attacked, until at last Lieut. Beaumont, Alexander Gray, sergeant-quartermaster captain of the sledge, and Frank Jones, stoker, were alone able to drag, the other four men having to be carried forward on the sledge in detachments, which necessitated always double and most frequently treble journeys over the rough and disheartening icy road; nevertheless, the gallant band struggled manfully onwards, thankful if they made one mile a day, but never losing heart; but Lieut. Beaumont's anxiety being intense lest relief should arrive too late to save the lives of the worst cases. Not arriving at Polaris Bay on the day expected, Lieut. Wyatt Rawson and Dr. Richard W. Coppinger, with Hans and the dog-sledge, started on June 22 to look for them, the two parties providentially meeting in Newman's Bay, twenty miles from the depôt. The following day Frank Jones being unable to drag any longer, walked; leaving the three officers and Alexander Gray to drag the four invalids, the dogs carrying on the provisions and equipage. On the 27th, Alexander Gray was obliged to give in, and the officers had to drag the sledge by themselves, Gray and Jones hobbling along as best they could. On the 28th, being within a day's march of the depôt with the dogs, the two worst cases were sent on in charge of Dr. Coppinger, and arrived at the end of the march, but I regret to state that Charles W. Paul, A.B., who joined the expedition from the 'Valorous' at Disco, at the last moment, died shortly after their arrival. The remainder of the party, helped by Hans and the dogs arrived at the depôt on July 1, and it being impossible to cross the strait and return to the 'Discovery' before the invalids were recruited, at once settled themselves down for a month's stay, those able to get about shooting game for the sufferers with such success that they obtained a daily ration of fresh meat. It was entirely due, under Providence, to the timely assistance dispatched by Lieut. Rawson, who, as senior officer at Polaris Bay, when there was not time to cross Hall's Basin and inform Capt. Stephenson of his apprehensions, acted promptly on his own authority and went to the relief of Lieut. Beaumont's party, that more casualties did not occur.

"After such details it is scarcely necessary for me to allude to the services of Lieut. Beaumont. The command of the Greenland sledges, entailing as it did the crossing and recrossing of Robeson Channel—which in 1872 remained in motion all the season—required even greater care and judgment than is always necessary in the leader of an Arctic sledge party. My confidence in Lieut. Beaumont, as expressed in my original orders to him, was fully borne out by his careful conduct of the party throughout this trying and most harassing march. He is a most judicious, determined, and intelligent leader, and as such I bring his services to the notice of their Lordships.

"Capt. Stephenson, by personal inspection having satisfied himself that the resources of the Polaris depôt were sufficient and appropriate for the subsistence of the men detached to the Greenland shore, although naturally anxious at their non-arrival on board the 'Discovery,' was not alarmed for their safety. On July 12, Lieut. Fulford, with two men and the dog-sledge, were despatched across Hall's Basin to Discovery Bay, and arrived there on the third day, having found the ice in motion on the west side of the channel, and experiencing much difficulty in effecting a landing., On the receipt of the news, Capt. Stephenson instantly started with a relief party, carrying medical comforts, and arrived at Polaris Bay on the 19th. On the following day the ice was in motion on both sides of the channel. On the 29th Capt. Stephenson, with Lieut. Rawson, Hans, and four able men, with two invalids who could walk, started with the dingy for Discovery Bay, and after a very wet journey they landed on the west shore on August 2, Lieut. Beaumont and Dr. Coppinger, with five strong men, being left for a few days longer in order to give the other two invalids further time to recruit. The whole party ultimately, re-crossed the strait, and arrived at Discovery Bay on August 14, having been absent from their ship 120 days, several of the party who had wintered on board of the 'Alert' having been absent since August 26 the previous year."

To the westward, the north coast of the American Arctic Archipelago was surveyed by Lieut. Aldrich.

Over the ice straight to the northward, Captain Markham, with Lieutenant Parr, proceeded to 83° 20', the highest latitude known to have been reached. Compared with the time it took, the number of miles made towards the North Pole appears small; but what were these miles? Contrary to what had been expected, no sheltering land existed to the northward, and the ice, by being pushed on the shore and screwed together, had been piled with an unevenness that appeared made to mock travellers in search of a sledge road. The blocks were flung up in hillocks and walls 30 feet high, and the floes often raised on edge, so as to render it impossible to cross them, and requiring passages to be hewn through them. The soft snow that filled the deep holes and ravines tended to complete the terrible difficulties met with. When Captain Markham started, with his brave associates, on the 3rd of April, the thermometer still showed -40°. During their extraordinary fatigues scurvy broke out, and attacked them one after another, so that the sick had to be dragged by the remaining healthy men. In returning to the vessel on the 5th of June, the latter were no longer sufficient to proceed farther with the sick. Lieutenant Parr then ran in advance to the ship, which he reached without resting in twenty-two hours. But before assistance could be brought the weakest of the sufferers had died.

During 142 days the sun was below the horizon—the longest winter night known. They also had the severest season that has been observed. During thirteen days the mean temperature was -58°, the greatest cold experienced -73° on the 'Alert,' and -70° on the 'Discovery.' On the 12th of March, Lieutenants Rawson and Egerton started in a dog-sledge with a Dane, N. C. Petersen. On the way the latter was overpowered by the cold. By the utmost efforts his companions succeeded in bringing him alive back to the vessel. But they had to amputate both his feet, and he died on the 14th of May. August 20th, both ships left Discovery Harbour on their return voyage.

Concluding Remarks

Hans Hendrik is now boatswain and labourer at the Greenland settlements. From the fees he received during his Arctic travelling life he has saved what, amongst his present comrades, may be considered a good deal of money, the interest on which makes a relatively considerable addition to his scanty wages. These servants having a pretty long time in the year for their private business, I suppose he continues the occupations by which he has earned both money and reputation, namely, sledging and seal-hunting,

I have said that the present account gives an idea of how a native Greenlander feels and thinks, and how he is able to express his sentiments. He left his home at the age of eighteen, and spent six years amongst the heathen natives in Smith's Sound. From his home he brought, besides some skill in reading and writing, his religious ideas, and especially that firm belief in a merciful God which has strengthened him in braving so many hardships. In other respects, notwithstanding his intercourse with the foreigners, he has maintained his nationality. I leave it to the reader himself to form an opinion of Hans and his countrymen, the Eskimo, from his own words.

In reading his pages I have asked myself how the narrator, if he did not keep a diary, could retain within his memory such an amount of details. In some places, especially in the last part, there are some indications of his having written notes during his travels; and his friend and late superior, Herr Rudolph, has told me that he received from Hans some papers in which he had described the country about Smith's Sound and its inhabitants. However, there is no doubt that the greater part of the narrative has been compiled from memory. For this reason, it is not to be wondered at if inaccuracies or confusion of details occur here and there; but, on the whole, such irregularities, if found, will be of very little weight.

The manuscript is written in tolerably plain and intelligible Greenlandish. But, as this is still a difficult language, as the writer is an unlearned man, and as I had nobody at hand to assist me, some words here and there remained inexplicable or doubtful, and some sentences unclear. These instances I have indicated by marks of interrogation. I have retained the curious spelling of foreign names, and the use of "thou" instead of "you," which has nothing corresponding in Eskimo speech. The denomination "Tuluk," English, here sometimes means English and American, sometimes the former in contrast to the latter. "Kavdlunak" properly signifies foreigners of in European race, whites, but here sometimes "Danes" opposition to other white men. Words such as "Tartikene" and "Tart Eise" show how imperfectly the author has understood English. I have guessed that they are formed out of "Doctor Kane" and "Doctor Hayes," but could not make out exactly what is meant with them. Explanations by myself are inserted in brackets: [ ].

For various reasons I have here and there omitted a few lines, and made the text a little less diffuse in form. But I have added nothing.

H. Rink.

Copenhagen, November 16th, 1877.

In looking over Director Rink's version, I have altered as little as possible. I thought it best to let Hans Hendrik write in the naïve way to be expected from such a child of nature.

G. S.

THE TALE OF MY TRAVELS TO THE HIGH NORTH

[The Author's Home, his Voyage with Dr. Kane, and settling down at Smith's Sound.]

To relate how the northern part of the big country came to be explored, I write this—I, Hans Hendrik, who first lived at Kekertarsuatsiak [Fiskernæs], belonging to the Germans [Moravians], but have now removed north, to Kangersuatsiak, belonging to Upernivik.

I was born in the German [missionary station of] Kekertarsuatsiak, which had three priests, and my father served the priests. He used to go to the other stations to lead divine service there on the great holidays. His name was Benjamin. My mother had the charge of the church lamps. Her name was Ernestine. She had come from the south, from Agdluitsok, from the end of the country [Cape Farewell]. Thus my dear parents at first lived far from each other, my father being an inhabitant of Kekertarsuatsiak, while my mother was born at the end of the land. My father's children by his first wife were seven altogether, three sons and four daughters; but by my mother he had five children, four sons and a female child, scarcely to be mentioned, because she died as an infant. I was next to the eldest, this female. I left my brothers still children, I was just growing a real kayaker, having got a few seals by harpooning. My next brother was only a small kayaker, his name was Simion. But next to him was Joel, who was trained up at Nouk to become a teacher, and he stayed there two years. Our youngest was named Nathaniel. The year after my father's death I departed, this first time joining Arctic travellers, an American sailing vessel.[1]

---
[1] Dr. Kane's Expedition, 1853.

I heard that they were looking for a native companion, and that his parents should have payment during his absence. Nobody being willing, I, Hans Hendrik, finally took a liking to join them, and I said I would go. The ship's Master tried to get one assistant more, but did not succeed. I went to inform my mother of my intention, and she gainsaid me and begged me not to join them. But I replied, "If no mischief happen me, I shall return, and I shall earn money for thee; but certainly I pity my dear younger brothers who have not grown food-winners as yet, especially the youngest, Nathaniel." At last we started, and when I left my countrymen and relatives, to be sure it was very disheartening. Still, I thought, if I do not perish I shall return. How strange! This was not to be fulfilled.

We left Kekertarsuatsiak and first landed at Maneetsok, but without anchoring, only trying to get some reindeer skins. Leaving this place, we met with a kayaker far out at sea, although a gale was blowing from the north. On asking him who he was, he said he was Amase from Nouk. The man I knew very well, as I had seen him often, a first-rate seal-catcher, now going to hunt reindeer north of Maneetsok. But proceeding northward we finally landed south of Upernivik, at Kangersuatsiak, and there remained pretty long, I have forgotten how many days. We went also to Upernivik in a boat, the ship sailing at the same time, but we reached it first.

At Upernivik we engaged as our companion a Kavdlunak [European], Carl Petersen, whom the natives called Naparsisortak [the new cooper]—a very good Kavdlunak he was, and he took care of me. When we came to Upernivik, I was invited to eat in the great house, as he lived in the merchant's house. He was married to the midwife, and she and their children all treated me kindly. Leaving Upernivik we went very far to the north, and we landed at a desert north of where people lived—it was not far from an abandoned winter station. When we had landed we fell in with the carcass of a white whale. Although pretty fresh, the birds had eaten the upper part of the skin, but the lower part was still untouched, and Naparsisortak and I got a store of *matak* [eatable skin]. Here we stayed, I think, for three days, and boiled the oil out of it.

Then again we started, but, meeting with heavy drift-ice, made the land, and came to a native house, the northernmost of the many wintering stations thereabouts: its name was Anoritok. As a strong southern breeze drove the ice a little from the shore, we went on thence. A strong gale was blowing; on one side of our craft the gunwale went under the sea. Finally, they could not help running against the ice, the ship's prow being lifted on its top. Not until it grew calm could they get off again. This was the first collision of our vessel with the ice. We continued between the shore-ice [ice-foot] and the drift-ice, and our ship at last ran aground. We then fastened it with a rope from the mast to the shore-ice, and when high water set in we proceeded behind the drift-ice and along the shore-ice. In this way we at length reached open water, inside of which we found a fine wintering-place, frequented by both hares and reindeer. But when I went out to shoot hares, I only got a few, sometimes one, sometimes two or three, and, although I searched for reindeer, I could not get sight of any. When winter set in we used to go ashore, two Tuluks [English] and I, to look for any sort of game; but we only met with hares and a few reindeer, the latter very shy, but the hares easy to get at.

In the first beginning of the winter a boat set out to the north, manned by nearly all hands. Only I and one Kavdlunak, with the cook and two sailors, remained. They stayed away for a long time, but I have forgotten how many days. At last, when the sea was frozen, they returned, walking over the ice, and having left the boat frozen up.

Then it really grew winter and dreadfully cold, and the sky speedily darkened. Never had I seen the dark season like this, to be sure it was awful, I thought we should have no daylight any more. I was seized with fright, and fell a weeping, I never in my life saw such darkness at noon time. As the darkness continued for three months, I really believed we should have no daylight more. However, finally it dawned, and brightness having set in, I used to go shooting hares. One day, when thus returning from a hare-hunt, I saw a crowd of people near the ship. Only think! they were the northern natives whom here I saw for the first time. On approaching the ship, two natives came running towards me. The foremost of them, when he had reached me, accosted me in a-very civil manner: "Art thou a native?" I answered: "Yes, I am a native." He: "Hast thou got any hares?" I: "Yes, I have got three hares." He: "With what sort of weapon hast thou got them?" I: "Look here, my gun." But when I spoke thus he did not comprehend, but examined it, how it was. I said: "It goes very far, taking hares, reindeer, ptarmigans, and Natsek-seals." On hearing this he started. When I arrived at the ship, I could hardly get along for the people, who would know what I said, but did not understand all my words. So, as I could scarcely get on board, the mate gave me a hand's turn, the natives, for mere civility, being unable to assist me.

When first I saw these people, whom I knew nothing about, and nobody had examined, I feared they might perhaps be murderers, as they lived apart from any Kavdlunak; but, on the contrary, they were harmless men. In the evening they went to sleep on board the vessel. The Tuluks offered them something to eat, bread and beef, and such like, with tea, but they did not relish them, they only tried some little bits. They said: "We cannot eat it;" and added, that they should like to have some hare-meat. But our Commander was careful of our hares. The next day, when they left us, our Master gave them wood, needles, iron, and matches, and they went off very thankful and cheerful.

After their departure the [frozen up] boat was found broken asunder ... and the sails in patches [?]—only think! a native had fallen in with it, and being unable to make out what this thing was amidst the ice [?] he had broken it into pieces. Our Commander, Kaine, grew angry, as he knew not who had done it. Later on, a native arrived on foot, named Majok. When I returned from hare-hunting I saw him shut up in the ship. The Master ordered me to examine him as to who had spoiled the boat. He said: "I don't know, I have not done it." The Master said he would shoot him if he did not confess. On hearing this I took fright; at once pitying him, and afraid to look at him, I uttered: "He says he will shoot thee if thou dost not tell." He replied: "I have not done it, I don't know it." Finally, unable to overcome him, they grew silent. Our Commander said to me, that he intended to shoot him. I answered: "What a pity!" We went to sleep, while he was kept prisoner. In the beginning of the night I heard a noise. I went out and saw him running off speedily. I wonder how he managed to get out, the hatch-way being very high. After his departure no natives made their appearance more, I think they were frightened.

Towards spring I began hunting Utoks [seals resting upon the ice], and usually got some, but I do not know how many, as I have not counted them. I also got four reindeer and several foxes. When they started on their first sledge journey, the sledge was dragged by men, and stayed away for many days. I afterwards set out to join them; I travelled by dog-sledge, and did not know the road, but had a map of the country. While still I went on, I saw a great bear coming straight up against me. I could not master my dogs any longer and was obliged to cling to the sledge lest I should fall off. But the bear took to flight, and I pursued and came up to him before I could loosen the dogs from the sledge. He was just going to reach an iceberg on the edge of an open water into which he dropped, whereupon my dogs stopped and some of them got free. I fired and hit his muzzle, but, as he crept upon the iceberg, I missed him.

I again gathered my dogs, and travelling farther I saw another bear, but I prevented my dogs from perceiving him; again I saw another blackish one, the natives yonder call them "A;" hunting them is very dangerous. I travelled and stopped at a large promontory, the distance from the ship being 100 English miles. Here I slept, and in the morning I started, travelled the whole day; in the evening I lay down to sleep upon the ice. The next morning I set out, and finally towards noon I met with the men I sought for. When I came to them they were upon an island and had got a bear. They were busy cooking, and regaled me with bear's meat.

Here I slept, and the next morning the others started in order to return, but I got a companion and continued northward. This companion was named Mister Morten. We passed a great glacier, proceeded the whole day, and did not reach land before next morning. Again we went on, and fell in with a large open water. We followed the shore-ice [ice-foot], but when this ceased we were obliged to stop. The next day we took a walk along the beach, carrying the dogs along with us, and saw a big bear running across a plain towards the hills. We called the dogs, and when they arrived I showed them the bear by running towards him, whereupon they stopped him. I then ran to them, but waited for my companion, in order to get a rest for my gun, whereupon I fired, and shortly after the bear tumbled over, quite dead. But the cub was killed by merely pelting it with stones.

We carried off the flesh of the cub, but only the skin of its mother. When we reached the shore we left our burdens and continued our trip to the north. Again we discovered a bear, but he escaped by plunging into the sea. Not far from a steep cliff I rested, waiting for my fellow; on his arrival we turned back and came to our stores. We had them dragged by the dogs, reached our tent, and being very fatigued got a good sleep. The next afternoon we ascended the hill to build up a beacon, showing the point we had reached, and we marked the rock.

The following morning we set out to return to the ship, and travelled, I do not know how many days, the surface of the ice becoming covered with much water. At last we reached the ship, open water being still four miles off, and remaining so the whole summer, as the ice did not break up. I used to catch hares, the country abounding with them. We also discovered bones of musk-oxen, and their crooked horns, but no living one.

In autumn, when the new ice was forming, some of our party set out for Upernivik [the nearest Danish settlement] in a boat. Naparsisortak joined them. I should have liked to do the same, but our Commander forbade me, saying they would be frozen up, and be unable to reach Upernivik. To be sure, as it had grown winter, Naparsisortak came back by sledge with one companion. They had left the others not so very badly off, but, beginning to run short of provisions, the natives used to send them meat as gifts.

In this way they came back, the natives carrying them to the ship by sledge. They looked very emaciated and were dreadfully voracious, but the Master bade them not to eat too greedily, fearing it might be hurtful. The next day his people, of their own accord, tried to cook meat in the native fashion, imitating the native lamps in their room in the ship. During this time I happened to go outside, and observed that our ship was taking fire. I shouted at once to those inside: "Our ship is on fire!" They then hurried out, and, drawing salt water from a hole, succeeded in quenching the flames. Our Master gave me many thanks for my quickness.

When spring came round, we started for the north on foot, dragging our sledge. After several days' travelling we put up our tent upon the ice. Having discovered many of their footprints I went out in search of bears. I fell in with a big one, who stood upright, but on my approach he squatted on the ground. While still drawing nearer I came to a large iceberg and heard a sniffing sound. I looked upward and discovered close above me another big one, who just emerged and disappeared. While I proceeded towards the first, which still was a little beyond rifle-shot, all of a sudden he took to galloping towards me. I fired at his head, but without hitting it properly. I was seized with fear when I had fired, and he tottered; but, before I had loaded my gun again, he went off stooping. I followed to finish him, but fired without hitting and lost my game.

As I was fatigued I rested on the top of an iceberg, but before I could manage to load my rifle I perceived something sniffing. Looking downward I saw another bear beneath me. I fired and he took to his heels, following the track of his companion. When I had rested I turned back to our tent and found all my companions asleep. In the morning we set out [for the bears, thinking that?] they were in a dying state. We discovered a large bear, but, although the ice was quite level, he disappeared … we heard their yelling … [?] as my rifle did not go off, I fired at him with my fowling' piece, but he scampered off, and then we saw no bears more.

In this place I only write down a little of what I ought to relate; so many years having elapsed I have forgotten so much. In winter, when daylight had commenced, they tried to go to the west side on foot, dragging their sledge, but, without reaching the shore, they turned back. I believe there were three Tuluks frostbitten. Naparsisortak and the Kavdlunak, carpenter, left their companions [and came back?]. We then directly started, taking with us the Kavdlunak, although he had just arrived and was perfectly exhausted. When we had eaten, he lay down upon the sledge, and we started dragging him, and having with us our Commander, Kain. We travelled the whole night, and finally reached them. Here we slept in the tent, feeling pretty cold, and when we awoke we returned, placing them upon the sledge. On the road we halted, almost unable to pull the sledge on account of the heavy load, four frostbitten men. But, proceeding on, we came to the vessel. A few days after our cook died from severe frostbites, also one of the sailors died. One of our officers, whose feet had been attacked by the frost, recovered; also the other officers recovered.

When spring set in the Master and I used to make excursions; he was very clever in not despising native food. Once, in the beginning of autumn, when we travelled by sledge over the new ice, we fell through. We carried my kayak with us, and the dogs having scented the footprints of a bear, we were unable to keep them back. On setting out upon the new ice the dogs fell through, and so did our Master, close by the sledge, only his head appearing above the water. I immediately stretched out my arm to give him help, but he forbade it lest I should also fall in. I rested upon the kayak, which was placed upon the sledge that had dropped into the water. My companion swam to the dogs and climbed up the ice by taking hold of the dogs. I succeeded in crawling from the sledge upon the ice. As we were unable to manage the sledge, we merely cut the thongs by which the dogs were tied, and took to running towards the ship. I feared my companion would have been frozen to death, but we succeeded in reaching the brig safely, by a forced run, I think, of about four miles.

One day also I started for a reindeer hunt. I returned to the ship after having stayed away for several days without having seen a living thing, excepting a hare. We also tried a trip across the sea to the west coast, but did not reach the large house [?], which was our goal, the heavy ice making it difficult of access. Also once I made an excursion with another to visit the natives, and, by the way, in kindling a fire to boil some water I chipped a stock and happened to wound my right hand, nearly cutting off my thumb. However, we proceeded in search of the houses, though ignorant of their site, and luckily we fell in with a sledge track, which led us to two houses. On our arrival we were received with kindness. When I held out my hand they pitied me, and bade us directly go inside; they would take care of the sledge and dogs, they said.

When we entered we found it very hot. I examined the house and found it wholly constructed of stones, even the very ceiling. Looking upwards, the large slabs seemed ready to fall down the next moment. We were instantly invited to eat and regaled with walrus-meat. They also ordered in some rotches [small auks], and when these were brought we found them quite fresh—only think! Now daylight had begun, and yet they had preserved during the whole winter the provisions gathered in autumn. To be sure those natives had an abundance of game, walruses, rotches, hares, and foxes. We even tasted white whale skin. These courteous people said, that the next day they intended to go walrus hunting, and would like to take my companion, the Tuluk, along with them. When I asked him, he complied, but I remained, as my hand ached dreadfully, being cut between the fore-finger and the thumb, so as almost to sever the latter. They returned with one walrus, my companion having got one of its big paws for his share.

We slept here two nights, and returned in one day to the vessel, my companion taking charge of the driver. On approaching the vessel, our Commander, who had observed that I did not drive the sledge, came towards us, and when, soon after, we had come on board, he said that he would cure me fairly, and added, that if I obeyed him the wound would soon be healed. He began curing me by cutting away a part of the inner flesh. To be sure it ached dreadfully when he took out the bad flesh, but it was soon healed.

After I had recovered, the Master and I repeated the visit, but on our way home he fell sick. Afterwards I once again called on the natives in order to take part in a walrus-hunt. I got one walrus. Eventually my companions began to think of abandoning their vessel and repairing to Upernivik. I did not believe they would be able to reach it. At the same time I happened to visit the natives in order to get hares. The day after I had come to them I set out for the chase in a gale from the north. A heavy squall suddenly carried me off, hurting me against the hard frozen snow. My native friends led me by the hands to the sledge, and carried me back to their houses, where I recovered during a stay of several days. As those men behaved so kindly towards me, I began to think of remaining with them. Spring came at last. When they were going, to abandon the ship, I proposed to our Commander to visit the natives in a boat, and, as he complied, I went off to them, and intended to come back. But they began persuading me to remain. My companions would never reach Upernivik, they said, and they would take me along with them when they removed.

A pity it was that our Master behaved with haughtiness towards his crew. Also, once he treated me in a similar way. The occasion was as follows:—I had cut the head off a reindeer-skin of my own catch, intending it for a sledge-seat. I went to the Kavdlunak [Dane], who was just taking a walk, and said to him: "The Master intends to shoot me for having cut the head from a reindeer-skin; that is the only reason." The Kavdlunak replied: "Don't be afraid, he will never shoot thee, I am going to say to him: we have another king." While he repaired to the ship, I stayed upon the ice, expecting I should be fired at, but perceived nothing at all. This Kavdlunak, on coming out again, said: "There is no reason for thee to be afraid, only remain with us, I will be thy protector."

From this moment I thought more and more of leaving them when they started for Upernivik. Finally I said to him [Kane] that I should like to visit the natives, and when he complied I departed. It was my intention to return, but I began to envy the natives with whom I stayed, who supplied themselves with all their wants and lived happily. At length I wholly attached myself to them, and followed them when they removed to the south. I got the man of highest standing among them as my foster-father, and when I had dwelled several winters with them, I began to think of taking a wife, although an unchristened one. First, I went a-wooing to a girl of good morals, but I gave her up, as her father said: "Take my sister." The latter was a widow and ill-reputed. Afterwards I got a sweetheart whom I resolved never to part with, but to keep as my wife in the country of the Christians. Since then she has been baptized and partaken of the Lord's Supper. But I was greatly delighted at taking along with me one of the unchristened, when I returned to a Kavdlunak settlement.

A Talk about the Unchristened Natives in the North.

In the days of yore their ancestors used to visit Upernivik, for which reason they still speak of "Southlanders." Those northern people had for their merchandise walrus-teeth, for which they got wood, whereas the Southlanders had wood to barter with. Their ancestors also possessed kayaks. One man, named Kassuk, when undertaking a journey southwards was told not to visit the Southlanders, because they used to kill their guests when they were going to depart. But he was an Angakok [wiseman, magician], and went off with his wife and children. He came to people who received him very kindly. But when invited to another house, the inmates accosted him as follows:—"Our neighbours used to kill their guests when they are going away, thou wilt not escape them." When they had spoken so, he again entered to his hosts, whom he found all asleep, except one young fellow who had to keep watch. This wretch he bewitched, making him fall asleep too, whereupon he went out and cut asunder the lashings of the other sledges and harnessed his dogs. Then he entered and

asked his wife to follow him directly; he was going to put to
the dogs, he said. So they set out over the ice before anybody
had come out; but, when they gained some distance from the
shore, a crowd of people made their appearance. They were
seen collecting their dogs, but on starting their sledges broke
down. Only one sledge continued to draw nearer,
accompanied by a dog of a frightful look. But, when this dog
had come abreast of them, he shot at him with his bow and,
hitting his side, killed him at once. Now, first, his dogs took to
run speedily, and left the pursuer behind. In this way the
traveller with his wife escaped, and from that time no other
journey has been undertaken to the south.

It is a great pity that people there in the north have no idea of
a Creator. Only by me were they informed about the Maker of
heaven and earth and everything else, of all animals, and even
of ourselves. Also, about His only begotten Son, who came in
the flesh for the sake of sinful men, for the purpose of saving
them, teaching them faith, and performing wonderful deeds
amongst them, and afterwards was killed on a wooden cross,
and arose from the dead on the third day, and will come
down again to judge the living and the dead. On hearing this
the Northlanders were rather frightened as to the destruction
of the world in their life as well as in their death [?]

How the Men yonder in the North subsist.

They pursue the white whales along the edge of the ice, using
five hunting bladders in connection with one line, but on the
big ice only one bladder. They get the seals which lie near
their breathing holes upon the ice, by creeping up to them and
harpooning them. They pursue the walrus by the aid of two
hunting lines, both ends of which are furnished with a
harpoon, and their spears are headed with a Tok [chisel]. As
soon as the line becomes tightened [pulled by the stricken
animal], they thrust this into the ice to hold.

They also catch seals by having many breathing holes at once occupied by men. One man then generally catches a great number, while the others only get a few, as the seals, when at the point of choking, have recourse to a few holes without leaving them.

Bears they kill by spearing them, after having brought them to bay by the dogs. They capture foxes in traps of four different descriptions. For hares they use nets made of sealskin thongs. For birds they also use an implement like a catcher.

Hunting seals at the breathing holes, during day, they call Marpok; during the night, Nirparpok. I also remember the following words of their language: sirlla pirnerkariarnarkark irnurk pirniarkark irgijarnarkark pirnirkark arllirnarkark narparpurk karparpurk pirniarllune arllarpurk sarnivarpurk irmirnirk narnurnijarpuk kirmursirgarpuk tirmirsarnut karllurpurk. Some of their words I have forgotten, as I left them after a few years' stay.

MY SECOND NORTHERN JOURNEY.

[Under Command of Hayes.]

Once during my stay at Kip John [Cape York?] in the beginning of autumn, we got sight of a ship. When drawing nearer they came close in from off us, lowered a boat and pulled towards us. When close by I recognised two gentlemen in the stern, the doctor and the astronomer, both of whom I knew were my friends. Before reaching the shore they shouted, calling me by name. I said: "It is I," on hearing which they were greatly pleased; I, too, was very glad to see these gentlemen, who liked me. They asked me: "*Ju* Hans Hendrik?" I answered: "Yes, I am it." They were also accompanied by three natives [of Greenland, Eskimos]. When they landed they said, that they wanted me alone to go with their vessel. I answered that I wished to take my wife along with me. They said: "Better let her stay behind, next winter thou canst go to fetch her." I replied: "I don't like to leave her, I pity her and her baby." They added: "Well, then, bring her." I said, likewise: "I will carry my tent with me." I then left the other tent to the parents of my wife.

We set out for the north, and while we sailed the wind freshened strongly, my dear wife soon fell sick, and so did I. After a little while we stopped to land. We went out after hares, because my wife was unable to live upon merely Kavdlunak food. I got two hares, the others got nothing. On coming on board I gave one of them to my wife, and earned many thanks from her. We stayed here for several days, and when we left we touched at the northernmost wintering station in an island called Pikiulek, but only on account of the heavy ice we met with. As the ice soon after spread, we went on, but were blocked again, and returned to a station called

Eta. Here we got sight of reindeer on the hills, and we landed towards evening, carrying our guns. I got no reindeer on account of my gun being a breech-loader, which I did not know how to handle, but one of the Kavdlunaks got one deer. As it became dark very soon, we returned to the ship, and went to fetch the reindeer the next day.

I felt very happy that now I had got something to hunt for. When I started in the morning I used to return at noon, sometimes with two, sometimes with four deer; in the afternoon I pursued the same sport. There were also a great many hares. In the beginning of winter, when the sea was frozen over, I used to go hunting alone. To be sure, I had three natives as travelling helps; but two of them were engaged with various labours on board the ship, and the third with some carpenter's work.

In the beginning of winter one of these natives turned a Kivigtok [fled from human society, to live alone up the country]. We were unable to make out what might have induced him to do so. The only thing we remembered he had uttered was—"What does J— say when he whispers in passing by me?" When he asked me this, I answered—"I don't know at all." Also of the others he inquired in the same way, but we were quite ignorant of what he meant. Once, when the sea was frozen, he went outside towards tea-time, as we supposed, without any particular purpose. But fancy! all of a sudden he had run away. After tea, on going out, I said to my wife—"Hast thou not seen Umarsuak?" She answered—"He went out after having handled his bag. I thought he was going to take his tea, but he said, "that is the only awkward thing, to understand neither Danish nor English."' As soon as he was missed, some of the crew lighted torches and set off in search of him. At last I found his foot prints going to the hills. I followed them and shouted to him, but got no answer. When I had reached the top of the hills, also there he had still run farther. But, as I grew exhausted, and his track disappeared in the darkness, I gave him up wholly and returned to the ship, where the rest had now arrived. Although we still kept up a faint hope of his return, he was quite lost, and his memory left a deep impression upon me, he being the only friend whom I loved like my brother.

In winter, just before Christmas, the astronomer and I undertook a journey by sledge to look for natives. We crossed [passed by?] the great glacier and travelled the whole day [of course only twilight, there being continual night] without meeting with any people. A strong wind sprang up from the north and caused a thick drifting of snow, while we made our snow-hut and went to sleep. On wakening the next day it still blew a gale and the snow drifting dreadfully, for which reason we resolved to return. While we proceeded homewards the ice began breaking up, so we were forced to go ashore and continue our drive over the beach-ice [ice-foot]. We arrived at a small firth, and crossed it, but, on trying to proceed by land on the other side, it proved impassable, and we were obliged to return to the ice again. On descending here my companion fell through the ice, which was nothing but a thick sheet of snow and water. I stooped, but was unable to seize him, it being very low tide. As a last resort I remembered a strap hanging on the sledge-poles, this I threw to him, and when he had tied it around his body, I pulled, but found it very difficult. At length I succeeded in drawing him up, but he was at the point of freezing to death, and now in the storm and drifting snow he took off his clothes and slipped into the sleeping-bag, whereupon I placed him upon the sledge and repaired to our last resting place.

Our road being very rough, I cried from despair for want of help; but I reached the snow-hut and brought him inside. I was, however, unable to kindle a fire, and was myself overpowered with cold. My companion grew still worse, although placed in the bearskin bag, but with nothing else than his shirt. By and by his breathing grew scarcer, and I, too, began to feel extremely cold, on account of now standing still after having perspired with exertion. During the whole night my friend still breathed, but he drew his breath at long intervals, and towards morning only very rarely. When, finally, I was at the point of freezing to death, I shut up the entrance with snow, and as the breaking up of the ice had rendered any near road to the ship impracticable, and the gale continued violently, I set out for the south in search of men, although I had a wide sea to cross. My nine dogs pulled quickly, and, by catching hold of the sledge-poles and running, I began to be revived. I passed by a great number of walrus breathing-holes. Before I could reach the opposite shore my dogs stopped from exhaustion. I was obliged to stay, and tried to sleep, but the cold kept me awake. At length I let loose the dogs, and set off on foot in search of a house I knew, thinking it might be inhabited. But by the way I tumbled down, and could only proceed by creeping. When I reached the house I found it empty; I made an attempt to sleep, and after a miserable slumber rose and fell down again, but at last succeeded in getting a little warmer by walking.

When I came to the sledge the hounds had almost consumed my store of provisions. I went back to the house, and then fell truly asleep. The next day I set out for another house I knew, but, before I reached it, my dogs again stopped from weariness. Then I threw myself down in front of a high cliff, awaiting my death. When here I lay prostrate, I uttered, sighing—"They say that some one on high watches over me, too;" and I added—"Have mercy on me, and save me, if possible, though I am a great sinner. My dear wife and child are in such a pitiful state—may I first be able to bring them to the land of the baptised!"

I also pronounced the following prayer:—

> "Jesu, lead me by the hand
> While I am here below,

Forsake me not.
If Thou dost not abide with me, I shall fall,
But near to Thee I am safe."

Thereafter I rose, and set off again on foot, after having loosened the dogs. I came to the house, and found it partly unroofed. However, by aid of the blubber I carried with me, I boiled some water, and slept. When I awakened I went to the sledge, put to the dogs, and proceeded. I crossed an inlet also on foot, because of the snow being too deep. I made for the station called Natsilivik, thinking there might be people. On approaching it I fell in with snow-bare ice, and sat down on the sledge again. I came to Natsilivik, but found also this place uninhabited, and thought that if the next station, Itivdlek, was so too, I should die. I cleared out the house ... my hounds ate [?] ... I slept here, and when I wakened I set off for Itivilek. On this road the ice was tolerably free of snow; but nevertheless, when I tried to sit down on the sledge, the dogs were not able to pull me along. I again took to walking, and when I had passed what is called Sermiarsuk, before I had discovered human footprints, the dogs got scent of them, and began running quickly.

I then sat down again, the dogs quite altered by following the footprints, and, how lucky! I discovered the light of a window. On coming nearer I fired my gun in order to warn them. At the same time people appeared, and it was to me as if I had reached my home. These folks were very kind and hospitable. When I entered the house and began to take off my clothes, the fox-skin of my jacket was as soft and moist as if newly flayed. My outer bearskin-trowsers were not so very wet. When I took off my hareskin-gaiters, they stuck to my stockings from being frozen together, and I could not get them off but by cutting open the boots. Had I used sealskin-gaiters I think I should have frozen to death. Here I stayed for many days, being unable to return alone. At last I took it into my head to fetch the parents of my wife from the farthest off wintering station, although I was greatly concerned about my dear wife and child, who lived solitary on board the vessel with no female companion. Our little daughter was still a baby, and they longed for me as their only support.

At last I travelled southward with two companions to a place called Agpat. We crossed [passed by?] a great glacier, and, after having slept once, arrived at Agpat, where I found my foster-brother, married to the elder sister of my wife, and now I came to him he was to me like a true brother. Here I remained for a great part of the winter, as it was difficult to make my return journey during the great darkness, finally, when daylight re-appeared, and the weather was fine, I sent off the husband of my wife's sister to convey my parents-in-law from the southernmost station, and I left him my dogs for the journey. After remaining away for many days he returned to my great pleasure, as now I got companions for my passage to the ship and felt very anxious. After a stay of three days we set out for the north, accompanied by other sledges, intended for a walrus-hunting station called Kejatarsuak. It took several days to reach this place, where we gave our dogs one day's rest. One sledge continued its course northward to the ship, so that when we arrived there, after three days' journey, the first sledge already had announced our coming.

But on board they would not believe in him who brought the first message; they thought he was telling untruths, and that I was dead. When they heard the news of my companion, the astronomer, having fallen through the sea and been frozen to death, they conceived suspicions, thinking our deaths might be owing to some crime committed by the natives, although, notwithstanding their being unbaptized, they abhor manslaughter. On my arrival I found my dear wife tolerably well; but I could not be happy since I left that friend of mine who had loved me so kindly, and who also, some winters before, when we spent three years together, had treated me with such goodness. Our Commander, Ese [Hayes], was gladdened by my arrival, as he had believed me to be lost. He enquired where I had left my friend. I replied: "On leaving him I covered him wholly with snow, now I will soon go to fetch his corpse." But he said: "When the days grow longer thou mayst go for it, better now first try to get some reindeer, we are longing for reindeer meat."

I then remained for several days to await a brighter season; the first day I went out shooting I got a large rein-buck. Afterwards I hunted every day, sometimes bringing home two deer, sometimes three. At last, when bright sunshine had begun, a sledge arrived which was engaged to accompany me. We also got the ship's mate for our companion. When we arrived we dug among the snow, and brought forth the dead man still enveloped in his bag. I feared the foxes might have eaten the body, but even the bag was quite untouched. We deposited him on my sledge, the mate followed with my comrade, and we came back to the ship in the evening. They brought the corpse into the Captain's cabin for him to thaw. The next day, when I saw our Commander, he said: "I thank thee for thy having taken care of him."

Now the bright season had set in, I gave all my time to hunting reindeer and hares. I also tried walrus-catching, and got a very large one, but being quite alone I merely fastened it [to the edge of the ice], after having killed it. The next day I returned with the sailors, an officer and two natives, to haul it up on the ice. But as it proved too heavy, we cut it up in the water, severing the head and a part of the flesh. What a size! I had never seen one like that during my sojourn there in the north. The natives as well as myself were used to catch walruses, but we never met with its match. When first I discovered it, it did not stir, and when it had dived I proceeded to the water's edge to watch it. At last I began to hear signs of its rising again, and as it emerged I harpooned it with the help of a very long line, furnished at each end with a harpoon. As it dived after being hit, I fastened the other end to the ice, and hauled in half the line; and when it emerged the next time I struck it with the other harpoon, so as to make my line become double, and then I killed it with the spear. When we had cut off its head and some of the flesh we returned.

After this I again took to hunting the reindeer; only once more I went in search of walruses, and got a small one, but then I chiefly followed reindeer-hunting, they being so plentiful. However, I happened to get nine hares also, and of the reindeer I killed too with my fowling-piece.

In the beginning of spring a man arrived by sledge, named Amalortok. He was accompanied by his son. They came in search of medical assistance, both of them suffering from a swollen throat, and at the same time to visit their relative, my mother-in-law. After a stay of several days they left us. On their arrival, the same men had told us that diseases of the throat and of the stomach were raging among the natives, of whom several had died. After their departure my mother-in-law also fell sick, got still worse and died. This was a hard blow to me, as she had behaved towards me like a tender mother. The amiability of these unbaptised people is to be wondered at; they are never false, but always loving towards each other.

While I was hunting reindeer in the beginning of spring, I once got nine deer, although I remained standing on the same spot. Later on, when the sledge parties prepared to go, our Master ordered me to remain, and collect reindeer for provisions. When some natives had arrived—three men, with their wives—the sledge party started, with Jensen, and the Captain, Mister Karmek, and Kemart—I have forgotten how many they were. Three days after some of them came back, having found it impossible to proceed. Some time after we went to fetch the boat, and brought it in one day. Later on I departed, carrying along with me my tent, and followed by the natives, whom the Master intended as my porters, to carry to the ship the haunches and legs of all the deer I happened to kill, whereas he asked us to eat only the saddles. During several weeks I collected reindeer; I did not count their number. At springtime my father-in-law let us, in order to visit his eldest daughter. A little before he started he was unable to walk, on account of his leg paining him, but as soon as he recovered he left with his son. This parting was very sad to me, as I could not expect to see him any more. All of us fell a-weeping. I had lived with him and his wife for several years, and they were to me like parents. Had she not died, both of them should have followed me to Upernivik.

We afterwards sailed, and touched at the island of Pikiulek, where our cable was injured, and we lost our anchor by a gale from the south. Our craft scraped the ground, but soon after came afloat again. I went ashore with some others, who got several reindeer; I only got a young rein-doe, a very small one. I also came to a place where the walruses use to creep up on the ice, and saw many of them, but only had at hand my gun. I shot four young ones of that year's brood, but their mothers rolled them into the sea. The next day I repaired to the same spot with my large hunting-line, followed by the sailors, and got four walruses. We brought them to the ship, hauled them up on the shore-ice, flayed them, and set to cooking at the same time. Here were also lots of eider-ducks, and we got plenty of them.

When we left this place we touched at the Westland, and, starting thence, steered for Upernivik. When we had anchored at Upernivik, and I came on deck the next day, the Master summoned me to the stern, and when I came to him, he thus accosted me: "People say that thou hast caused Pele to run away, him who turned Kivigtok during the winter?" I answered: "I have not made him run away, who says that I have done so?" ... Thereupon the crew collected around the Master. One of them, Carl Emil Olsvig, passing by me, said:

"Don't fear, I will help thee and speak in thy favour." When they had finished, Doctor Rudolph [the trade-agent, who had come on board] said: "Farewell! to-morrow thou and thy wife come ashore and stay at Upernivik, I will take care of thee."

The next day we landed, to our great delight, and were lodged in the house of Julius. The Captain came too and proposed to convey me to Fiskernæs, but I answered that I wanted to remain at Upernivik, and he consented. Our Master now appeared to be quite changed, full of love towards me, and liking me as he did in former years. He gave me a gun, which I took, but another rifle, which likewise he offered, I refused, as it did not shoot straight. He also added tea and other provisions and pork, and a wage-sum of 72 *dollars* [Danish?]. Doctor Rudolph proposed that I should settle down at Augpilagtok; the trader of this outpost would take me there; no station was equal to Augpilagtok, he said. I spent three winters there, one at Upernivik, and one at Kingitok. At last I removed to Kangersuatsiak, and was engaged by the trade as a labourer. While living at Augpilagtok I had work at Upernivik every summer. At Kingitok I taught the children, the clergyman having sent me there for this purpose. In spring I went to Upernivik, when the merchant wanted me to work there. Doctor Rudolph was so kind towards me and my little wife that I could not refuse him. Finally, when taken in regular Government employ, I settled down at Kangersuatsiak.

MY THIRD JOURNEY TO THE NORTH

[With the 'Polaris.']

While I was at Kangersuatsiak, a post arrived from Upernivik informing me that Americans had reached there, and bringing two letters for me—one from the merchant, the other from Kekertarsuatsiak [Fiskernæs]. I found the letter from Upernivik to contain in substance:—The American Arctic travellers wanted to have me in their service; a boat was coming to fetch us, my wife and children being invited to go with me. They had also taken on board with them a native from the Westland [coast opposite to Greenland], with his wife and foster-daughter, and the American, Mister Morta, a friend of mine, whom I knew from my first voyage, when he was steward, and I also had him for my companion on my sledge-run to the far north. When I opened the other letter, it was from Kekertarsuatsiak, written by my brother, and informed me:—Since Simeon had perished in kayak they were alone, and their condition very sad indeed, and last year another person had perished in kayak; but, thank God! all of us have a future haven, howsoever we fare by land or by sea, and our days are numbered. When I opened the third letter, I noticed it was not written to me, it being written in Danish. I merely had a look at the writing, and directly folded the paper again.

While I was reading over my letters, I heard shouts of "A boat!" and that it was white, and of a foreign appearance. Already knowing the Americans were coming, I went up the hill, and on the way met with the Assistant Trader. When I had informed him that the Americans were at hand to fetch me, he said: "Thou wilt not be allowed to join them, as thou art in debt." I answered that I was ignorant of my debt, and added: "By mistake I have broken the sealing-wax of the third letter; I did not understand it, as it was in Danish; I will deliver it to thee." When I had gone to fetch it, and given it to him, he grew very angry, saying: "Why hast thou torn off the seal?" I answered: "In ignorance I did so;" whereupon he ordered the Guardian to be called. The Guardian asked: "What does he mean?" I answered: "As I did not know a letter addressed to him, I broke it open, believing it was intended for me." He communicated this to the Assistant, who then grew silent.

The boat having landed, the Assistant Trader said: "The merchant wants thee to join them." A little while after the ship's Mate, Mister Tarsta, said: "What pay dost thou want per month?" I answered: "Ten *dollars*" [Danish?]. He: "It is too little, is it not?" I said: "Twenty-five." He again: "It is too little." At last, as I did not demand any more, he asked: "Will fifty be sufficient?" I replied: "Yes, that will do." He added: "Art thou willing to perform sailor's work on board or not?" I agreed so to do, with the exception of going aloft. When I had spoken thus, he was satisfied, and said that we were to start the next day.

In the evening, Mister Tarsta remarked: "I will sleep in your tent." When we lay down to sleep, Mister Tarsta did the same, and the next morning my wife, having boiled coffee and cooked meat, regaled him with a cup of coffee, but with no meat, thinking he did not like it, although this was quite a mistake [?]. In the morning we departed. It was very sad when our friends came down to the beach to bid us farewell, as some of them were as dear to me as kinsfolk. It was then beautiful weather, and we sailed briskly along, with a fresh easterly breeze, I taking care of the sheet.

When we approached Upernivik, and came aboard the ship, my late companion, Mister Morta, rejoiced at seeing me again. The Master of the ship, Ull [Hall], was very kind and friendly, as were the natives from the Westland, John and his wife. We stayed one day, and, proceeding northward, touched at Tasiusak, whence in three days we came to where lived the natives.

The next day, when we made for their northernmost settlement, Eta, but still pretty far from it, I said to my comrade, the Westlander: "Yonder I lost my companion." He answered me: "Don't say so;" whereupon I grew silent. When our Master had gone away, I asked him: "Why didst thou forbid my speaking?" He answered: "The officers do not like to hear it, as they believe thou caused his death." I replied: "That never struck me when I spoke to thee; is it possible that any baptised man should be able to think of such a thing?" My comrade did not utter a word more.

Proceeding, and passing the northernmost station, Anoritouk, we steered for the Westland. On the way my wife fell ill from stitch, and grew still worse. At last John's wife said: "I think she ought to be cupped [?]; I should like to try it. It is a very dangerous case; I have seen many such; some got worse, some improved." I agreed, and she went to fetch her blood-letting instrument, and put it on her back, saying: "Now she will grow very hot, and when she feels difficulty in breathing, use this fan;" whereupon she sat down at her side. As she had said, the heat came on, and she felt half-choked, but fanning brought her breathing all right again.

In this state my wife became delirious, and said: "Itakavsak will come over us"[?]. She now grew very talkative, and continued: "Now listen to something. I remember that last summer a *Kivigtok* came up to me, whilethou [Hans] wast on the journey to Upernivik with blubber. In the evening I went to fetch water. When I came to the large stone, a man approached, he who had disappeared last year; ye remember, he went out shooting, and did not return. When he drew near he said: 'I come to thee as thou art a stranger [belonging to the far-off Northlanders?]. Last year I ran away because I was in love with Nepisak's daughter, but they [her relatives] would not have it, when I was joking with her. She also was rather fond of me, and as they forbade me marrying her, I went away. I was also tired of needlework, my sister and mother being unwilling to sew my clothes; for this reason I fled from men. Last spring I met with my father in Satut, and also his brother, and last autumn I came upon the dogs of the Assistant at Itivdlinguak. When I was going to slip off, I practised it thus: I drove to the coast opposite to us, and coming to a small opening, I deposited my gun and my mittens on the ice, and pushed my sledge, which was loaded with two seals, into the water. Then I went ashore, and observed my two younger brothers running up. As soon as they came to the sledge they began lamenting [believing I had perished], and turned back; but after a little while a swans of people hastened up, and having reached the sledge, began searching in the water, although I was standing upon the hill and looking down upon them. When they started to go home, and my brothers and relatives cried out for me, I joined in their wailing. At the same time I heard some of them say: "It sounds as if somebody was crying up yonder." When I perceived this I stopped my crying.' The Kivigtok, after having finished this, grew silent. I went off to make it known; but I don't know how it was, when I met another person who fetched water, I had forgotten what I had to say."

My wife continued her tale thus: "Also once afterwards, while Hans was gone to Upernivik, and I slept with my children in the tent, a heavy shower pouring down, I heard a voice outside of some one wanting to come in. But among our dogs sitting oh their haunches in the entrance there was one of brownish colour [?], and this was the reason why he could not enter." [She added that] When she had taken a book, and placed it in the inner part of the tent, he finally went away [?].

What here my wife related was certainly true. On the day when that man disappeared, in the evening there was made inquiry for him, and people said that he had not returned. His two younger brothers had found the sledge floating in the water, and his mittens and gun placed on the ice. But they searched in vain for the owner. I believe also I can remember that when we started to go home from that spot, we heard a wailing; and when we came home, it was mentioned also by others that they had heard a sound as if somebody was crying on the hills. This is the end of my report on my wife's vision.

While we went on far towards the north along the Westland, we also arrived off the furthest point reached by Mister Morta and me on our sledge-journey. When my comrade questioned me about it, I remarked: "There we got two bears." That country was named Kip An Tusen. The great Master of the ship, Ull, during that time, was very praiseworthy; he scarcely allowed himself any sleep. We proceeded still farther towards the north, between our country and the Westland, meeting with pack ice of an amazing height. At length we were wholly stopped; the ice closed us up and drove us back, I don't know how many miles. At last we began discharging the cargo on the ice, fearing we should be wrecked. But the next day we took it in again, as the ice spread away. We made for our country [the Greenland coast], and when Captain Ull first went ashore in a boat, they shouted *"Hurra!"* at having reached our wintering harbour.

Here we saw many wild geese upon the land, and the next day I went ashore with one man more, each getting four of them. There were lots of them, but they were very shy. One day, when we took a walk over the hills, we discovered footprints of musk-oxen not quite recent. On coming on board we reported this to our commander, who grew very glad to hear there were oxen in this place.

On September the 8th four of us, namely, the. Doctor, Mister Tarsta, John, and I, set off sledging in search of oxen. We travelled the whole day, and took a rest after a journey of 8 [English?] miles, without having seen any live thing. The next day, I, with Mister Tarsta and John, went out on foot, leaving the dogs; only the Doctor remained, to look for some fine stones. We searched the whole day, but only discovered footmarks of oxen, not quite fresh. On our return I said to the others: "I will go this way," and parted from them. When I came to the tent, the others had arrived, also without having found anything. The next day we went in another direction, walking across the eastern hills, behind which we found an inlet covered with ice for a long way, the road to it leading over an extensive plain. We returned to our tent in the evening. Again off in the morning in another direction. We soon returned. While, towards evening, we were busy collecting stones to secure the tent, my comrade came up, whispering: "Look there up the hill; may that be a fox?" I turned round to look, and saw a big beast, whose legs were hardly visible on account of its hairy coat. To people who have not seen such before, it looks rather terrible. When it had stared at us for a while, it drew nearer. The dogs began yelling, and tore asunder their harness, but on coming up it attacked them, raising a cloud of dust about itself. As we could not get quite near, I fired at a distance, as did my comrade, but still it continued to pursue the dogs. We discharged our pieces I don't know how many times before it fell. We had it dragged to the tent by the dogs, cut it up, and set to cooking. The flesh was similar to that of reindeer; the nape of the neck was very big. Next day, back to the ship, and our arrival gave the Master much pleasure.

Afterwards we used to go shooting. My comrade, the Westlander, and I sometimes caught seals; in this way we roamed about together. In October our Captain made up his mind to sledge, and we set off in one overland. We found the snow very deep, and stopped after 4 miles, made a snow-hut, and lay down to sleep. When we awoke, the Master asked me to return to the ship, and fetch one sledge more. So I set off, and after having stayed a short while on board, as they invited me to eat, I returned with two sledges and more dogs. After our arrival we proceeded again, my comrade with the Commander, and I with Mister Tarsta. We halted, and made a snow-hut, continued two days more, came to the sea, and took rest on the shore ice. Here the Master said: "To-morrow ye may go shooting, while Mister Tarsta and I push on afoot."

We started for the open water. I shot a seal which emerged through the new ice, and towards evening my comrade fired after another in the same way, but missed. When it appeared again I hit it, but the current drifted it to sea. Towards evening, we returned, and later on our officers joined us. They reported that they had found a large hole in the. Glacier [?]… and made steps in the ice with a knife … lest they should slip down [?] When they heard of the seals I had lost, they were vexed, as they had taken a great liking to seal-flesh while staying for several years in the Westland. The Captain said: "To-morrow we will return to the ship."

We went to sleep, but kept quiet next morning, on account of a gale from the south. The day after we set off, reached our snow-hut, and slept there. After having travelled one day more, we again reposed. When we started thence, we at length got sight of our craft. We stopped for a moment to take our meal, agreeably excited by this view, whereupon all of us, shouting, hastened on board.

After our arrival, in the beginning of the evening, my comrade entered our room, saying: "Our Father is sick." I asked: "Is he very ill?" "Yes; he is, very sick." I rejoined: "But to-day he did not appear ill." Next morning, on meeting my comrade, I asked: "How is our Commander?" He answered. "Just the same." We went out shooting; each of us killed a seal, but my comrade did not get hold of his. For some days we roamed over the hills, but without seeing anything, except at times some tracks, of hares. We also tried shooting at the water's edge, and sometimes saw seals, but found them too wild. Several days later our Captain grew better. My wife and children stayed every day in their room, according to his wish; but on returning in the evening, I used, to go visiting. The Master, who was now quickly recovering, once said, that as now he felt well he should like to go out the next day. But when we had gone to sleep, we were wakened by my comrade, who entered, saying: "Our parent is very sick." I replied: "Why, he was lately quite well!" When, thereafter, I went to see him, he was weak, and kept feeling the left side of his chest. My comrade accosted him, but got no answer. On seeing that he could not live, we said to each other: "When he is no more, how shall we fare, for he is our only protector?" When again we had gone to sleep, after midnight, he who kept watch knocked at the door, opened it, and said: "The Master is dead." When he had spoken thus, I said to my wife: "He says our Commander is dead; how will it go with us hereafter, as he was the only one who took care of us? What mournful news, that he who loved us so kindly lives no longer!"

Later on, we put our Master into his grave. When the days shortened, the ice broke up, and drifted us with it, our ship dragging its anchor in a heavy gale. My comrade and I both advised the Master who now had the command to drop the second anchor. When he had ordered the sailors to drop it, the vessel stopped. But then again we drifted towards a large iceberg [?] He ordered the sailors to ascend the berg [?], and fasten a warp. As none of them was willing, my comrade and I mounted the berg. It blew dreadfully, and on stepping down we found our faces frostbitten. The vessel was now fastened to the berg.

This was the first danger we incurred after the death of the Captain, who was so watchful and careful of us. While he lived we caught seals, I don't know how many, and geese, and I got one Ugsuk [a very large seal] and one musk-ox. When we were frozen up close to the iceberg, he collected the crew, and exhorted them not to be anxious, as they had two native providers, if we should be wrecked there. When the ice had formed, but before it had any thickness, my comrade said to the Master [?] that the iceberg would capsize. He did not mind, saying there was no fear. But when the ice grew thicker, our craft began to incline at low water, and to right itself at high water. At length she began to draw water, and it looked as if the iceberg would break and swallow her.

At the end of the dark season, when daylight began, we went out shooting. We also saw some seals, but they were too shy. Although at neap tides thin ice was forming, we, nevertheless, had open water during the whole of the winter, sometimes distant 3 miles from us. When it became light, it was about as far off. During the darkness the open water was close to the ship; when daylight appeared it withdrew.

When we had got out of the dark season, we used to walk up the hills, but there was no trace of oxen. We only got some hares, I have forgotten how many.

Once, when the sun had re-appeared, I heard that I was to be punished in man-of-war fashion. The sailors informed me: "To-morrow, at nine, thou wilt be tied and brought down to the smithy. Mister Tarsta will come to fetch thee after nine o'clock. Take breakfast without fear; if thou art afraid, thou wilt be treated ill." When I heard this, I pitied my wife and my little children. The next morning, when we rose, towards breakfast time, my wife, our daughter, and I fell a-weeping. Our little son asked: "Why do ye cry?" From pity we were unable to give an answer. However, they brought us our breakfast, and, though without appetite, we had just begun eating, when we heard a-knocking at the door. It was opened, and Mister Tarsta, with a smiling look, made his appearance, and accosted us: "*Godmorgen*, are ye eating?"—whereupon, still smiling, he petted our children and left us, and a heavy stone was removed from my heart. I also thanked God, who had shown mercy to a poor little creature. However, since the Captain's death, Mister Morta, and Mister Tarsta, and Mister Blaine, and the Clergyman, and some of the sailors were pretty kind towards me.

When it was daylight, and my comrade and I went out shooting, I said to him: "Why have they this horrible custom?" [corporal punishment]. My comrade answered: "Thine and my custom is to be ashamed of [the white men despise us?]. I cannot speak about it." When he had said this, I was long silent; then I rejoined: "Although in some measure I have an idea of that custom, I am nevertheless unable to understand it quite [?] Twice [I have seen it?], first in *Tartikene* [with Doctor Kane?], the second time in *Tart Eise*" [with Doctor Hayes?]. Whereupon he put in: "Yes, on board the man-of-war ships they are unable to abandon that custom. I should like to revenge [?] a relative of mine, a Westlander, who was treated in the same way. And likewise I saw a Tuluk, a very handsome youth indeed [exposed to it?]. But we poor natives must be very careful with regard to ourselves." When he had ended, I added: "Now first I begin to understand that detestable custom. While I was young, I would not believe it; I did not think it was said in earnest, and, moreover, did not understand English sufficiently. When I return, I believe I will not go with the Americans more. But should the English want to engage me, I will go with them. People say they are better." In this way we gossiped on what we had observed during our journeys.

After the sun had returned, I got one seal by *Maupok*-catch. [watching the breathing-holes]. My comrade got none. He and the Doctor and the Clergyman made an excursion by sledge;

but three days after they came back without the Doctor, to repair the sledge, which had been damaged. They also wanted me to join them with one sledge more. The morning after their arrival we set off, my comrade intending to drive with the Doctor, whom we reached in the evening, lodged in his snow-hut. When we had slept, we started, advanced the whole day, and stopped in the evening facing an open water. In the morning my comrade and I went shooting from the edge of the ice. I killed two seals, but only got hold of one, the other being carried off by the current. My comrade shot one, but lost it in the same way. When we came back it became pretty hot in our snow-hut. We set to cooking meat, and furnished our camp amply with blubber; it was in the native fashion, with a long wick. Next day, walked to the south over the shore-ice [ice-foot]. The Doctor said they intended to go to the southern cape, and asked when I thought we could be back again. I said: "To-morrow;" but he objected: "We want to be back to-night."

We started, and wandered the whole day, and not before midnight had we reached our goal. Here we rested for awhile, started again, and then came back quite exhausted. It was also a thirsty trip; not before our return could we get anything to eat or drink, and we had to lie on the snow. The Doctor and the Clergyman began sleeping. I was nearly doing the same, but, fearing the cold might be dangerous, I wakened them after a short slumber. Oh, how cheerful it was when we gained our sledges, and could sit down upon them! We met with John, who had waited for us, and, when he had slept, had gone to look for us and bring us refreshments. He said: "I was very anxious about you; I thought a bear had devoured you, or ye had fallen into the sea; my fears were many. I think now ye are nearly starving; here I have brought bread and *Panike* [?], but no cooking-pot." When we had rested and eaten, we went on again; where there was no shore-ice we

MY THIRD JOURNEY TO THE NORTH

were obliged to carry the sledge. On the other side [?] we came to my comrade's dogs, and thence to our snow-hut.

When we had slept here, the Doctor and John set out in a sledge for the interior of the firth, while the Clergyman and I remained. For some hours I went shooting on the edge of the open water, but without success, as no seals were to be seen. When I came back, the Doctor and John were still missing. It grew evening, and not before late did they return, having killed a bear—a small female one. They brought it entire upon the sledge, and then first it was cut up.

The following day we remained to let our dogs rest. But then we started on our return. We travelled the whole day, and a part of the night too, and arrived on board the ship the next morning at eight o'clock.

While the sea in front of us was covered with ice, we used to walk over it, and watched two breathing holes, but could not get at the seals. At last we began talking about going to look for musk-oxen. In the beginning of May we set out by sledge, John and I alone. We travelled the whole day, following the valleys between the mountains, but without discovering a living creature; we only perceived some tracks of hares. In the evening we stopped, and made a snow-hut, and had a lamp after native fashion, with blubber and a cooking pot. We had a supply of dog's food and *Pamikes* [?], containing fruits and sugar. Our eight dogs, as well as ourselves, were well provided for.

Here we stayed two days. We roamed about and searched in vain for oxen. On the third day we descended from the hills, and removed to a firth called Ingeverman Bay, on the border of which we built our snow-hut on the shore-ice. Next day, off sledging into the firth. When we came to an iceberg we brought up; and while my comrade examined the land through a spyglass, I remained on the sledge. But when he exclaimed: "Ah! just look in the glass, I see a small musk-ox moving over the snow." I went to look, and observed something like a big stone moving. "We searched for more, and discovered nine others, in the same direction. Hastened in pursuit; but they were very far off, on the top of the hills. Over the ice we drove pretty quickly, our hounds beginning to grow excited. When we reached the land, my comrade took the lead, to hinder the dogs from making any noise. It was a long time before we came up to them. Then we loosened our team. When the beasts observed them they collected in a circle, with their horns pointing outward. My comrade said to me: "Their way is, that when they show fight to the dogs they retire, but then suddenly advance, and then again go back, sharpening their horns by rubbing them against the ground." When he had spoken thus, we fired, and some of them fell. But, lo! a large bull drew back, and then suddenly rushed forward in pursuit of the hounds. It was awful to see the snow rising about him, like a cloud. Again he retired, facing the dogs.

We continued shooting until we had finished them all off, and had got nine oxen in one day. When we began skinning them, my comrade said: "When thou observest an oxen quite free from blood, fill it up with snow; we shall then get water for drinking, we are much in need of water." As soon as I saw one of them with no blood inside, I filled it with snow, and after a little it had thawed, and we were provided with drinking water. When we had carried all our game down the hills, we went to our snow-hut, and had a good meal of beef. The following morning we fetched more of the flesh, but leaving a part of it. Again we went for more, until we had collected the whole in our snow-hut. Thereupon we took one day's rest, and then we repaired to the ship. We travelled the whole day, and arrived the next morning.

On board they had been very anxious, and thought we were lost. Consequently, when, on approaching, we shouted that we had got nine musk-oxen, they were delighted. After three days' stay we went off in two sledges, I with Mister Tarsta, and my comrade with Mister Maje. In one day we reached the snow-hut, and discovered on our arrival one musk-ox more. We went to sleep, and next morning walked up the hills, I with Mister Tarsta and John with Mister Maje. We proceeded towards the spot where the day before we had observed the beast. On approaching and spying after it, we got sight of one ox, and, looking farther, I discovered eight more of the same kind. I proceeded on foot towards the first, telling my companion, whom I left behind, that when I made a sign with my hand he might approach. When I came close to the beast I waved my hand, whereupon he drove towards me in the sledge. As he disappeared behind a hill, I grew impatient, and went to look after him. I had left my gun upon the sledge. When he came up I said: "I will fetch my gun." When I had brought it and came to him, he had already hit the game several times, but not before I joined him did we succeed in killing it. After having taken out its entrails, we went in pursuit of the other eight.

We found them browsing separately, but on seeing the dogs they crowded together, forming a ring. We began firing at them, and succeeded in killing them all. While we skinned them my companion broke silence: "Now I will go cook beef and boil coffee, while thou continuest skinning. When I have finished I will assist thee." When the meal was prepared, and we had eaten our meat, we finished skinning and cutting up, loaded our sledge, and returned before our companions had arrived. After awhile they appeared. When we had slept, we set off in both sledges to fetch our store of flesh. We ended our job, bringing it down altogether. Next morning we returned to the ship.

Although we had our large supply of flesh to bring home, my comrade and I remained on board for several days to rest. Then we went off again, I, as usual, with Mister Tarsta, and my comrade with Mister Maje. When we had gone half-way we stopped and put up our tent, and thence reached our late encampment, with the stores, by the following day. After a day's rest we started on a trip to the end of this firth, to explore its interior. We found the firth headed by a glacier, and its interior also covered with perpetual ice, on which rested an immense number of stones. We made a sketch of the interior, and turned seawards again. On the road we raised our tent on the ice, and at last reached our late encampment. Next morning we crossed to the opposite shore of the firth, and ascended a hill, where Mister Maje at noon observed the sun, and reported that we were ... higher [latitude?] than the ship. When we started from this spot we travelled a long way, turned back in the afternoon, and built a snow-hut in the evening. Again advancing, we descended to the ice, and fell in with the traces of two oxen. Following these, we found a cow with her calf, and killed both. At length we came back to our camping-place, and thence to the ship.

After a stay of two days we went off again, I as before with Mister Tarsta and my comrade with one of the sailors. When we had stopped in the evening and began cooking, my comrade went up the hills and discovered two oxen. We hunted them on our sledges and killed them. After one day's stay we returned to the vessel.

Again, after several days, we got help to bring home the flesh. On reaching our encampment we obtained two oxen more. We returned on board, and, later, we two alone went out in one sledge, but got no more venison. Coming back we learned the sailors had been more fortunate, they had carried a tent with them, and brought in two oxen. After this our excursions ceased as the land grew too moist. In June, however, my comrade and I got some Utoks [seals upon the ice].

About this time there began to be a chance of boats passing northwards, although the heavy pack which incessantly came drifting from that quarter rendered it very difficult. However, two boats and provisions were transported over the ice a distance of 3 miles. My comrade asked whether we were to join the party, but the Master replied: "No, better be near the vessel; we must keep you as our purveyors." The skiffs then started, but during the following night one of them returned. The officer in charge, Mister Tarsta, reported that they were wrecked. The ice had crushed the other boat, he said. Some of the luggage was lost, but the crew were all saved. Now they had come to get a skinboat and provisions. When the ice opened they departed, but in the beginning of July one of the men arrived with the message that 25 miles off they had been unable to proceed any farther. They were closed up and cut off from the vessel.

Early in August we left the harbour with our ship, and tried to reach them, but were stopped by ice. We fired the ship's gun to warn them, but they could not hear it, whereupon we turned back. In the evening a heavy gale sprang up from the north. I went to sleep early in the morning, but was wakened at eight o'clock, and ordered to bring a letter to them and recall them. First they pulled me ashore in a place where I could not land, on account of the shore-ice offering no footing. Then they brought me to an ice-bare beach, but here I had to ascend a high rock; I was obliged to climb a precipice, which appeared quite inaccessible. Imagine my joy when I had reached the top! I set forward, travelling the whole day, and when at length I sighted the ice, and discovered two men at the edge of the water, to be sure I felt easy. I proceeded towards them, although with difficulty, from the ice being inundated. When I came to them, and inquired about the others, I was informed they were half-a-mile off. There were Mister Tarsta and the Doctor, and, as I had letters to the Doctor, I repaired thither. When I had found them, I said, that the other day we had started in the ship, but had been obliged to turn back, and then had tried to warn them by firing a gun, and that now we were waiting for them to set out on our home voyage. Mister Tarsta rejoined: "The ice has been quite impassable." When I had slept there, I started in the evening with the Doctor; we went the whole night, and reached the vessel the following day at noon.

In August all the others likewise returned, while the ship was still retarded by the ice. Every day I went up the hills to watch the state of the ice, as the Master had ordered me to give warning when it was going to spread. On the 11th of August, in the night, my dear wife was delivered. What a happy result! After a little while she was restored, and, both being well, I went up the hill to watch the ice.

I found it had began to clear off. Speedily went down and reported to the Master: "The ice has spread; we ought to be off to-day." On hearing this he directly went ashore, and after a little while returned, whereupon he made ready to start. It was afternoon on the 12th of August when we set out. But after two days we stuck in the pack, and were brought down with it towards the south. While thus we were blocked, my comrade and I caught seals every day, and then began collecting a store of unskinned seals. At the same time while the ship rested immovable, they put up a tent on the ice, and filled it with bread. When we were off Kap Allikisat, a gale sprang up from the south. It was a pitch dark night, when the ice began moving northward, and the floes were jammed and pushed over each other. At last our ship began to crack terribly from their pressure. I thought she would be crushed.

On perceiving this we brought our wives and children down upon the ice, and hurried to fetch all our little luggage, and remove the whole to a short distance from the ship. Then the ice broke up close to the vessel, and her cables broke; but in the awful darkness we could only just hear the voices on board, and when the craft was going adrift we believed she was on the point of sinking. Here we were left, ten men, our wives and children, and the Tuluks, making nineteen in all, and having two boats, no boat remaining with the ship. When the others drifted from us we thought they had gone to the bottom, while we ourselves were in the most miserable state of sadness and tears.

But especially I pitied my poor little wife and her children in the terrible snow-storm. I began thinking: "Have I searched for [?] this myself, by travelling to the north? But no! we have a merciful Providence to watch over us." At length our children fell asleep, while we covered them with ox-hides in the frightful snow-drift. At dawn our Commander Tarsta said he would make for the land with the men, as soon as their meal was done. When they had cooked and got their breakfasts they set off towards an island called Pikiulek, but before they could reach the shore they were stopped by new ice.

About this time we sighted the ship, which was approaching us, to our great joy. They steamed on, and I believed they would have observed us, but suddenly they turned, a heavy squall from the north coming on at the same time.

When our Tuluk companions were going to make for the land, they asked us to follow them, but my comrade and I preferred to stay behind, knowing that they could not get to shore. The cook also kept us company, saying that he found it pitiful to abandon us. Those who had tried to land returned after a while, not having succeeded. The north wind blew furiously, and the heavy seas threw us towards the Westland. Suddenly the ice on which we dwelled parted, and we were separated from the tent which contained our store of bread. When the ice touched the Westland it stopped, and packed together all around us. Here we made a snow hut. My comrade went out sledging, and how lucky!—he caught sight of the tent. Directly we started, dragging a boat to fetch some bread. At the tent we filled the boat with bread, and drew it over the ice to our camping place. When we left our wives and children I was afraid a bear would devour them, now I was consoled to see them unhurt, and after our arrival we had a good meal. Since we left the ship this was the first time we ate sufficiently.

The following day we deliberated whether we should remove to the floe where stood the tent, as it was very large and might serve us for an island during the winter. We resolved to proceed and first brought thither one of the boats, loaded with bread and luggage, whereupon we filled the other in the same way. My wife and daughter loaded the sledge with our little properties and pulled it, my wife carrying the baby in her hood. Our son was seven years of age, our youngest daughter four, and these poor things walked over the rough ice, my wife and daughter pulling the sledge, and I assisting those who dragged the boat—a sad sight. When they were going to be left behind, I told my wife I should return to her. When we had brought the boat to our new camping-place, I went back, followed by one of the sailors, and, finding my little daughter Sophie Elisabeth very tired, we placed her on the sledge, and more men came to help us. When we had finished our removal, we turned the boat over, I and my family going to sleep under it, while the Tuluks were lodged in the tent, and the Westlanders made a snow-hut for themselves.

The next day we built a snow-hut in the middle of the ice-floe. Fancy! this was to be our settlement for the whole winter. One day we rested; then my comrade and I went out sledging towards the land. On approaching it we fell in with new ice; I remained to look for breathing-holes, while my comrade proceeded towards the shore. I found some holes, and heard the sound of breathing; but as the ice was covered with snow, I could not get at the seals [which were scared by the noise]. My comrade had been on shore, and told me he had seen footprints of hares and foxes.

When we returned, we made up our minds to remove to the land the following day. We also drove in another direction, but without discovering anything. Next morning we tried to go shorewards, but our island, the ice-floe, began moving. It drifted seawards, consequently we turned back, and now we continued to be carried off incessantly in a southern direction throughout the winter. After some time we caught sight of land, but by-and-by lost it again. Every day my dear comrade, the Westlander John, and I went out hunting. In this way once he succeeded in getting a seal. What a joy, when we had a meal of flesh, and our lamps became supplied with blubber! Afterwards I again got a seal, a small one; I killed it at one shot. Wonderful indeed, we were so blessed with seals for our support, and that we so continued the whole winter.

Once, when we were out shooting, I fell through, having both legs under water. My comrade asked: "Art thou wet?" I answered: "No, I did not get wet." When we had tried shooting we returned, but quite near to our encampment a strong northern gale suddenly overtook us, and made both of us lose our way. The snow drifted terribly. As I was tired with walking, I stopped. Looking up towards the sky, I perceived many stars. Thereupon I proceeded, but came to a broad crack, and on going back, I fell in with the open sea. Now I thought my last day was come. I considered the miserable position of my dear wife and children, on a piece of ice in mid-ocean. Then I pronounced my prayer:—

> "Jesu, lead me by the hand,
> While I am here below;
> Forsake me not.
> If Thou dost not abide with me, I shall fall;
> But near to Thee I am safe."

When I had finished these words I ascended a heap of ice-blocks, and discovered a star rising a little above the surface of the ice. But it was my comrade, who had lighted a torch, and pointed it all round from the highest part of the uneven ice. I went down in the direction of what I saw; but on my road I again fell in with a fissure, turned, and went on, but again discovered something like a light. I moved forward, examining it, but was again stopped by the break. While here, some people were heard approaching, and when they came close they shouted: "Art thou Hans?" I answered: "Yes." Whereupon they said: "We had nearly fired at thee, believing it was a bear." I answered: "Never more I had reason to be thankful to anybody than to you, as I was quite unable to make out whither I had to go." When we came home I found my wife and children had been most sorrowful, but I thanked the merciful Providence on high.

While we drifted in this way throughout the winter, my comrade and I frequently got a seal. Our lamps were never out for want of oil. When sometimes our supply was almost consumed, one of us used to catch. Just before Christmas, each of us took a seal. How delightful, that our lamps were well supplied for Christmas! During Yule we finished all the provisions we had, except the bread; but we were consoled by knowing that daylight was near.

When the sun reappeared, we fell in with a great many black guillemots. Of course we also availed ourselves of them, as we were well off for guns—I had four myself, namely, three rifles and one double-barrelled fowling-piece. And we had plenty of shot. These articles I and my comrade John had taken care to provide ourselves with when we left the ship. At first we only threw them down upon ice, then we brought them some distance from the ship. We could, therefore, afford to shoot guillemots.

Although the sun again shone, no land could be seen, and it was truly appalling to think that our Tuluk companions and our wives and children would probably starve. However, we were taken care of by Providence, and the whole winter were supplied with seals. While still we lived on our island of ice, we fell in with bladder-nose and saddle-back seals, and they gave us a good supply of food.

As we advanced far south, we had a heavy swell, and, in a pitch dark night, the floe, our refuge, split in two. At length the whole of it was broken up all around our snow-huts. When we rose in the morning, and I went outside, the sea had gone down, and the ice upon which stood our house had dwindled down to a little round piece. Wonderful! There must be an All-merciful Father.

Some days after, when we had gone to sleep, we heard a gun fired. I went out and saw that a bear had been hit and had fallen. My comrade exclaimed: "We have got a big bear; how cheerful, we shall now have bear's-flesh!"

When we came still farther south the ice appeared more dispersed, and at last we made up our minds to go in search of land, although none at all was in sight. At the same time, we again met the heavy swell. We started in the boat, which was heavily laden. For some days we pushed on pretty well. When the seas came rolling they looked as if they were going to swallow us up, for which reason, at intervals, we landed on ice-floes. At length we made out land.

Again we rested upon a piece of ice. Daring the night a heavy sea came on; we slept with our children in the boat, while the others used the tent. As the sea still rose higher, it began washing over our place of sojourn. They were obliged to remove the tent, placing it upon the top of an ice-hillock, whereupon all of us had to keep hold of the boat. The children were placed in it, the women assisted us. When the sea began to move the boat, we all kept hold of the gunwales; the breakers looked as if they would engulph us. We exerted ourselves to the utmost each time when the sea began lifting us, whereas when it retired we pushed the boat to remove it to windward, because there was a danger of our being washed down into the sea to leeward. We did not stop until we had brought the skiff close to the edge of the ice. But now the sea reached the tent which was placed on the hillock. To be sure it was awful! whenever the waves washed over us we were in water up to the waist, while at the same time we clung to the gun-wale, and all the while one heard nothing but exclamations: "Now use all your strength."

Towards morning the sea had abated, and when it grew light we discovered that some smaller floes were less exposed to the swell. I spoke with my comrade about removing to one of these, and our Commander Tarsta agreed. We put the boat into the water, loaded it, and went to a smaller ice-floe, which we found much better as it was not washed over.

As the sea grew calmer we pushed on. Seals were plentiful; we had no want of meat; and we used to take our rest on the floes. One night it happened that the ice which served us for our camping place parted between the boat on which I slept and the tent. I jumped out to the other side, while that piece on which the boat was placed moved off quickly with Mister Maje who was seated in the boat, and we were separated from it by the water. Our Master asked the sailors to make a boat [raft] out of a piece of ice, and try to reach it, but they refused. We never had felt so distressed as at this moment, when we had lost our boat. At last I said to my comrade: "However, we must try to get at it." Each of us then formed an *Umiardluk* [literally, a bad boat] out of apiece of ice, and in this way we passed to the other fragment. As now we were three men, we could manage to put the boat into the water. But when on doing so, it sank forward, Mister Maje fell into the sea. My comrade jumped into the boat at the same moment, and pulled him up; I, being unable to follow, remained standing on the ice. When they had taken me along with them, we proceeded towards the others; but meanwhile the ice had screwed together, and we stood still. We three men alone then hauled up the boat [?]. At this time night fell, and our companion who had been in the sea, and now was lying in the boat, was like to freeze to death. I said to my comrade that if he remained so he would really die; if he could walk about, it would be better. I had witnessed such a case before. When I had spoken thus, we asked him to rise, saying, that if he remained, he would perish. The first time he rose, he tumbled down; but, after having walked for a long time, he recovered. At day-break we discovered our friends close by, and the ice joined together. When first they had examined the road, they came to us and assisted us to drag the boat over to them.

When we had started from this place, we were soon stopped by the pack, and no live thing was to be seen. We began to be in need of provisions. We had no seal-flesh left, and the next day our small stock of bread was to be shared out. In the night I had just fallen asleep, as I was to have my turn of the watch, when I was wakened by hearing people speaking about a bear. Rising up, I saw a bear walking towards us. I said to the others that they must lie down near the boat, imitating seals [?], while my comrade and I went towards the bear, who alternately sank and reappeared behind the ice hillocks. We waited until he came close up to us, whereupon my comrade gave him a shot, and I finished him off. Thereupon the others joined us to drag him to the boat. How wonderfully did Providence bring us through the winter, and give us supplies! At length we were off the remotest part of the Westland, whither the ice had brought us since last year: we left the ship in the far north. We were now near the country of the Tuluks without having suffered any real misfortune. Before we had finished the last of our bear's-flesh the field opened, and we began catching seals, and sighted land, and when we proceeded towards it we fell in with a ship.

Once in the afternoon, while still making for the land, we discovered a vessel steaming northwards. We tried to follow it; but night fell, and we stopped at the ice. At the same time there rose a dense mist. During the night we showed two lights near the boat, making them pretty large, that people on board might observe us. After midnight I went to sleep, when the others had risen. Towards morning I was awakened by hearing them talking about "ship;" and when I got up I saw it emerging from the fog. I directly set off in my kayak, and when I came to them they questioned me: "Who are ye?" I answered: *"Nord Polen mut Polaris Bebeles"* [peoples?]. Then furthermore they asked: "How do ye do?" I answered: *"Captain Ull Diet"* whereupon they said: "Where's the ship?" I answered: "Last year we left it." On hearing this they said to me: "Just follow a little alongside the ship, we will soon stop her."

When we had come up to my companions, they lay to, to take them on board. I was the first who set foot on deck, then followed the others; and when all had come on board it was as if we were ashore. The Master of the ship and the crew altogether were exceedingly kind to us, and pitied us who had spent the whole winter, with our little children, on a piece of ice. They gave us tobacco and pipes, and, before all, a good meal. Their Master, from mere kindness, was like a kinsman to us.

When the mist cleared we discovered another vessel close by, steering towards flocks of seals which lay upon the ice. Then the sailors in a body went seal hunting. We were informed that the other ship now had a full cargo of seals, and intended to start on their return voyage. We delivered letters to them concerning our rescue, but our ship had first to complete its catch before returning.

In a few days we were ready. We then proceeded two days, I believe, towards land and three days to reach the homestead of the Captain. First we stopped in front of a hamlet where they said some fishermen lived. Then we repaired to the master's place. Here for the first time we saw horses used for draught, a very strange thing indeed to me and my wife and children, though we had heard talk about it. The day after we went ashore. The Master invited us *Inuks* [Eskimo] with our children to a meal: he was very loving towards us. We stayed one day, on the third we left for another place called Nevland. We travelled one day and reached it in the evening.

Here resides an American consul [?]. We went ashore with the Captain to take a view of the place: he looked so pleasantly at each of us when he took us along with him. He brought us to a house of the most enormous size. On the ground floor there were many rooms each with its waiters, I and my wife and children were lodged in the middlemost floor, the crew on the uppermost. Our countrymen the Westlanders came to another house. Our Master Tarsta and Mister Maje stayed with us in the large house, and we had a luxurious living, no wants whatever, and were presented with dresses. Every day we went out driving with horses, to keep us healthy. Our youngest son, however, from having dwelled throughout the winter in such a severe cold could not properly get strength. We also wished to have him christened, and asked for some one who could baptize him. Then a Clergyman, only think! a Kavdlunak came to us and performed the rite in our room, some few people attending.

After some days a ship came from America to fetch us. When we were going to embark, the man who had to take care of us questioned me: "How is thy intention; art thou willing to settle down in America, or dost thou prefer to return to thy country?" I answered: "I should like to return to my home land when there is an opportunity." He replied: "Thou wilt come to thy country all right."

We embarked in the vessel, a man-of-war. We Inuks were lodged aft in the Captain's room; we felt very bashful, but were encouraged by the great kindness which the officers showed us. In the evening some soldiers near the ship sang very nicely. The following day we started for America. We travelled, I have forgotten how many days, and came to

Washington, where the Chief of America lives. Next morning they asked us to go ashore, first him who had been our Master during our stay on the ice, after him Mister Maje, after him John, and lastly me and my wife and children.

When we had come ashore the officer [Secretary of State?] next, in command to the Chief of America, questioned me: "From what sort of disease did Ull die?" I answered: "I did not know his sickness quite, but it was similar to stitch; first he improved for a while, but then he had a relapse and suddenly died. When I went to look at him, he tried to grasp his left side. This is all that I have to say." He also asked me: "What dost thou prefer, to settle down here, or to return home?" I answered: "I wish to return to Upernivik when I can get an opportunity." He replied: "Thou wilt soon return;" and he added: "Come to my house, all of you, to-night."

In the evening, three carriages came to fetch us. We drove between numbers of houses and came to many doors, and the doors were opened, and there were also many soldiers and black men. We passed by the house of the Chief, which was of an amazing size, and came to that of his subordinate. On arriving at the entrance we went out of the carriage, ascended big steps and came into a large house-passage. Then again we went down a broad staircase below the surface of the earth, where we entered a large room. Here he regaled us, and treated us very politely. He also gave us some images, his own portrait and that of the chief. When we had finished our meal we returned to the ship.

The following day we left for a place farther to the north and a little cooler, as they were anxious for us on account of the heat. First we touched at Boston, stopping, I believe, for one day, then we landed in New York. Here we stayed several days wondering at the crowd of masts, the large buildings, and heaps of people. While on board, we went on deck in the evening to have a view of the many vessels that lay

MY THIRD JOURNEY TO THE NORTH

intermingled, illuminated by thousands of lamps, some of them furnished with red, some with blue, and some with yellow glasses; and their whistles were heard, some very shrill, some less so. To be sure it was an amazing view, the lights of the houses glittering like so many stars. At last I began to think on the creation of the world, and I said to my comrade: "How wonderful that all these people subsist from the trifle that the soil produces; behold the numberless houses, the charming shores yonder, and this calm sea, how inviting!"

After some stay we were ready to travel by land in a carriage to a sound which we were to cross. Before we had seated ourselves I saw a man coming towards us with a friendly face. On drawing nearer I recognized Mister Trillis, the little doctor who was amongst the people of [the?] Tartikene [Doctor-Kane?]. He said: "Dost thou remember me?" I answered: "Ah, Mister Trillis, I think." He then said: "Yes, indeed it is I, come to see me." I rejoined: "We are just going to start for the opposite side of the sound." He behaved very kind towards us. We seated ourselves in the carriage, went off and entered a large building, inside of which we stopped, still sitting in the carriage, and waiting for some who were to go with us. I thought we were in the house, when my wife said to me: "Just look behind." We looked round and noticed that what was behind us was now sea. Now first I perceived that we were in a ship. We crossed and landed on the other side, still without leaving the vehicle, and we continued our drive among many houses, until we arrived at a ship by which we had to sail. Only then we stepped out of the carriage and embarked for the interior of the firth.

The vessel was very large, with two funnels. In the evening at tea-time we entered the eating room, where we found many gentlemen sitting at a table; we were placed at another table, and felt very shy. When we had finished we thanked the great gentleman who treated us so politely. On going to bed we were told by one of the officers who had care of us, to sleep without undressing, and to look well after our luggage. Early before it grew light we were to land and go in the great waggon. During the night we were wakened, and carrying our things we went into a vehicle, which took us to where we had to await the railroad train [literally: steamer by land]. We entered a large house, where we were regaled in the eating room. The Captain of the ship guided us thither, his name was Christian. He intended to bring us to his father-in-law, who had been an officer in the army, but now had retired from old age. When we entered the large building he asked us what we should like to eat. We got some fresh meat. Soon a train came in, but they said that ours would be in later. When those people went out who were to start with the first, they looked like a crowd of church goers, on account of their number.

Then our train arrived, and we took seats in it. When we had started and looked at the ground, it appeared like a river, making us dizzy, and the trembling of the carriage might give you headache. In this way we proceeded and whenever we approached houses they gave warning by making big whistle sound, and on arriving at the houses they rung a bell, and we stopped for a little while.

By the way we entered a long cave through the earth, used as a road, and soon after we emerged from it again. At length we reached our goal and entered a large mansion, in which numbers of people crowded together. It was almost stifling, I said jokingly to them "Get out of the way." On the same occasion we lost our companions, but on coming outside we found each other again. Here a carriage drawn by horses arrived with an old man, who now was to take care of us, the father-in-law of him who first had attended us. He invited us to seat ourselves and drive to the country of the farmers [cultivators], to be lodged in an uninhabited house. We came to a very fine land, over which lay spread the cultivators' houses. Numbers of horses, oxen, and sheep were seen all around, as if they were without owners, and the habitations scattered here and there offered a beautiful view.

The Captain, who brought us to his father-in-law, had said: "I will come back to fetch you," and, bidding us farewell, he had left us. His father-in-law likewise was named Christian. He used to come and look after us; every third day he brought us provisions. Whenever he came his first words were: "My children, how do ye do?" He was a very pleasant man indeed. When he left us, he inquired what sort of food we should like. He also used to send us lumps of ice. To be sure, it was very hot; in calm weather we did not take a walk without being provided with fans.

During our stay here in the country of the farmers, lodged in the empty house, we used to get eggs and milk from another house, where there lived people, our foster-parent having allowed us to fetch them there. We also frequently had visitors, who came in large carriages. Sometimes, at noon, when we were going to have our dinner, and people crowded in, we felt embarrassed. However, they were all very kind, and they used to give our children small coins and sweets.

A clergyman also came to see us, from a place in the neighbourhood. On leaving us he proposed to my wife and me to attend service in the church next Sunday, following the western road. On Sunday we set off, with our little son, but we missed our road, and after a tiresome walk returned. The following Sunday we went off again, choosing another road, and, passing by the farm-houses, we reached the church. When we came up to the people, who stood waiting outside, a very friendly man accosted us, saying: "When I enter the church, ye must follow me." So we did, and he invited us to sit at his side. This man was the schoolmaster: he started the hymns, whose tunes sounded wonderfully fine. When the parson had done, he sat down in his place, to examine his pupils, that people who were present might hear them. The cleverest first answered, whereupon the others joined in a chorus. When the service had ended, and we left, another man invited us to his house, this being nearer. The parson and the schoolmaster drove home, I don't know how far off; but he whose house was close by the church regaled us, and showed us its surroundings.

We spent a great part of the summer in the farming country. In July we heard that two ships were going to leave for Upernivik. One of them, which had arrived from the country of the Tuluks, had run aground, but was able to get off again. Afterwards we heard that this was to carry us. The assistant of the great gentleman came to bring us to our embarking-place, New York. We and my comrade and his family, who were to remain in the country in a seaport, now left. We travelled first by railroad train, then in a carriage, then crossed a sound in a steamer, proceeded a short distance on foot, and then again by steamboat reached New York. On arriving here, the ship which was to convey us was taking in cargo, for which reason we stayed here several days.

My comrade and I went to call on a gentleman, an acquaintance of his, who had wintered five times in his country, the Westland. We passed a road between a great many houses. Being ignorant of where his friend lived, we entered a merchant's shop to make inquiries. The merchant, on seeing the portrait of my comrade's friend, and, being asked as to his home, showed us the road. He said that when we had passed two places where the road divided, we were to mount a carriage drawn by horses, and there get further information. He added that another carriage was soon coming which might take us, and he gave a piece of paper with something written upon it to my comrade. When we went outside, the carriage had just passed. I ran after it and made signs with my hand, whereupon they stopped to take us. We only drove a short way, when they pulled up, and we entered to another shopkeeper. He likewise instructed us, and, when we left him his two daughters followed, to show us the road to the carriage. When it came up we made a sign and showed the paper, whereupon they allowed us to take a seat. We drove northward, and when we stopped they pointed out a large house which was our destination. We went towards it, and came to a railed-off place, opened a gate, and entered. We found it overgrown with beautiful trees, laden with fruits of every description. We followed a large road without seeing people. We came to the door and knocked. As nobody heard us we opened it, and came into a large house-passage, at the end of which we knocked at a door, and were asked to come inside.

We found one man here, and my comrade recognised his friend who had spent five winters with him in the Westland. When he asked about me, he was informed that I was the North Pole traveller, Hans Hendrik. On hearing this he knew very well from report that I had been much spoken about. We all agreed in lamenting the death of our Captain [Hall]. He also knew him well, from having roamed about with him during five years in the Westland. Towards evening he said: "Anything ye should wish to demand, any amount of money, I will give you." My comrade and I, after having deliberated, replied, that we should like to have some cigars. Of course, I also liked money, but could not use it now, as it could not be bartered in other countries [so I tried to give a courteous answer?]. When we left, the gentleman followed us to a shopkeeper, whom he asked to give us cigars and tobacco. We got four boxes each, and one more filled with cut tobacco, to be divided between us.

In the evening we returned on foot, as now we knew the road, and on the way frequently met armed men who had to watch the road, and to whom we applied for information. A few days after we left in the ship. We touched at a place called Newfoundland, to take in coals. On the third day we started for Upernivik, but on the road touched at Kekertarsuak, where also we got coals; I believe it took us three days. From thence, on the third day, we reached Upernivik. Thus ended this voyage.

MY FOURTH VOYAGE TO THE NORTH, WHEN I WAS ENGAGED BY THE TULUKS

[The English Expedition under Nares.]

Once I set out in a boat to fetch blubber from Southern Upernivik [an outpost about 40 miles south of Upernivik]. When we had departed we soon fell in with southerly wind, and therefore made for a harbour called Ingiudlertok. Ascending a small hill and looking seawards, I sighted two ships off the place where we were, and made out that they were Arctic explorers. I went down and set off again for our destined place. The wind was not very strong, but the current took us to the north. The breeze abating, we took down the sails and rowed back to the harbour. I again went up the hill with the Kavdlunak cooper. He questioned me about the road to Southern Upernivik. I pointed out to him where one had to go either by crossing the hills or by coasting. I thought he had asked me the question without any particular purpose. When we came down to the boat he asked me for a walking-stick. I gave him the longest tiller, thinking he intended to walk to Southern Upernivik, though he did not mention it. However, we waited for him till towards evening, when I went up the hill, and spied him with the glass. In the evening a southern gale sprang up, with heavy rain and a foaming sea. I feared we should have been driven ashore, as, for want of a spot to fasten a rope, we only rode at anchor. I did not sleep the whole night. My men tried to sleep, but could not, on account of the rain. In the morning, when the wind abated, I said to my crew: "I will turn back now. If we remain here, we shall have nothing to eat. However, I know that the assistant will reprove me, thinking I have done so for the sake of the Tuluks." We reefed sails, took up the anchor, and started. We made a good headway, with a favourable wind. On approaching our settlement [the outpost *Pröven*, belonging to Upernivik], we put out the reefs, as the wind lessened.

On entering the harbour we found that the Tuluk vessels had arrived. When we were going to anchor, the assistant [outpost-trader] came down, I feared to scold me; but on the contrary, he accosted me very friendly: "I am glad thou hast returned, otherwise the Tuluks would have gone to fetch thee. Thou art to follow the northern explorers, taking Matak along with thee." When I heard this I reluctantly agreed. I went up to my house to take my best clothes, before I was ready they shouted outside: "The assistant wants thee." When I came out to him, I found there the Tuluk officers who had come to ask me whether I was willing to go with them or not. At the same time the assistant gave me a letter, by which I understood that I was to go with them. Consequently when they asked me whether I was willing, I complied. They also talked about a companion for me. I said I should like, as I went along, to pick up my wife's brother who lived near Kip John [Cape York?]. I believed him to be a good hand at building snow huts. But as I was now going to depart, I pitied my wife and my little children who were so attached to me, especially my only son who would not cease crying, as he preferred me to his mother. I said to the master of the ship, that I should like to take my little son and my daughter Augustina along with me to Upernivik, where they were to remain. Thereupon I left Kangersuatsiak, making my fourth visit to the north, with the Tuluks.[2] When we put to sea and I looked at the people on shore, through the spy glass, I discovered my little daughter, Sophia Elizabeth, lying prostrate on the top of a big stone and staring at us. It was a sad sight which made me shed tears from pity. But I felt consoled by thinking that if no mischief should happen me or her, we should meet again. I also got sight of my wife standing amongst the crowd and looking

---

[2] Hans was on board the 'Discovery,' Captain Stephenson.

after us, I said to myself with a sigh: "May I return to them in good health."

We arrived at Upernivik in the morning and again left in the afternoon j my two little children followed me to the beach. We made our way behind the islands. East of Kingitok we stopped for a little while, as the other ship had run oh a reef, but soon got off again. North of Kingitok two kayakers left us, who had followed us as pilots from Upernivik. We continued to proceed towards the north, until the country where people live came to sight, called Kip John by the Tuluks and Ivnanganek by the natives. When we were off this coast, we parted company with the other ship, to visit the native settlements, and try to find the man I wished to take along with me.

When we landed I observed footprints of men in the snow, and supposed they were those of people living in the eastern settlement. When they had made fast the boat to the ice, I went over land with an officer and a Kavdlunak cooper, whom the Tuluks had engaged. While we walked over the hills, I observed a sledge standing on the ice near the waters' edge. We made for the eastern hamlet, but when we arrived we found the houses empty, and only some sealskins spread to dry. The sledge which we saw had first been driving in an easterly direction, but when they discovered the vessel they turned towards her. When they came close to the ship, I asked them: "Do ye know me?" They answered: "Yes, dost thou recognise us?" I rejoined: "I only recognise that old man amongst you." Who the others were I could first make out when they had stated their names, as I knew them well when they were children. I questioned them: "Is Augina well?" They answered: "Yes, pretty well." "Where is he?" "On that island yonder." When I heard this I said to the Tuluks: "They say that he stays yonder on the island." The commander then wanted the natives to go and fetch him by sledge, but they said they did not like on account of the great distance. Just when we arrived here, the Tuluks had caught a narwhal: they gave the natives some of its skin to eat, and some biscuits. They did not care very much for the bread, but greatly preferred the *Matak* (whale-skin).

We started and proceeded along the edge of the ice towards the island, but stopped before we came close to it. I then went quick by sledge with two companions, an officer and the Kavdlunak cooper. By the way we discovered two sledges driving north of us. I instantly directed our course towards them, and coming up asked: "Where is Augina?" They answered: "Yonder on the island, we come thence making a long circuit on account of a crack." I repeated this to my companions, who replied that I might go thither: they intended to return to the ship, I answered: "If I go, I shall not be back before to-morrow, the captain has ordered us to make haste." When I made this objection, they agreed that we should only return to the ship. These people also soon recognized me, though I did not them, before they told me their names, as I had only seen them in their childhood. The old ones I knew quite well. We repaired to the vessel followed by the natives; after having searched in vain for him whom I wished to engage.

Starting from this place we continued our course northwards, and touched at Eta. Here we went up the firth to hunt reindeer, and came to the head of it. We had the Captain with us. His steward asked me to wait a little as he wanted to go with me. When I went off with the master I fell ill from stitch. I got sight of a hare, and when we drew nearer there were two, of which the master shot one, and I the other. When we had ascended the top of the hills and proceeded towards the glacier, my illness increased, and I said to the Captain that I should like to turn back, to which he agreed. In returning by the way, I picked up the hares. My stitch grew worse; when I came to our camping place they were going to cook. I had a severe cough, shivered from cold, and felt very ill. Our

MY FOURTH VOYAGE TO THE NORTH, WHEN I WAS ENGAGED BY THE TULUKS

companions came back with one reindeer. When we had finished cooking and eating we returned on board. The following day we left. I now grew very ill, and thought my life was near its end, although we were unable to know the number of our days. The doctor, however, gave me medicine, but it was of no avail, and I could eat nothing. At length he cupped me on my back, and this took effect, I began to improve.

We now went north, coasting the Westland, while our ships sometimes were stopped by ice. Whenever it retired from the shore we went on, and when it grew very bad we made fast to an iceberg. In this way we reached the narrowest part of the sound between our country and the Westland.

On the 18th of August I caught my first seal, a Natsek (firth seal). On the 22nd, in the night, we arrived at our wintering station. When in the morning I came on deck, they told me that on board the other vessel they had got a musk-ox, and I believe three walruses, and that they had seen seals. It was a great joy to know that here, there was something to hunt. I was also informed that the other steamer was to proceed farther, while we had to winter in this place. When the others were going to start, I went ashore to accompany them in my kayak some distance and bid farewell to my comrade, the other native. I stopped at the side of the ship and talked with him, whereupon I went ashore in search of musk-oxen. I ascended a hill and got sight of a large one. I approached, taking care he should not see me. While he went up another hill, I came close above him. Just as he saw me he took to running, but stopped to stare at me, whereupon I fired, but at the same moment he rushed on me as if unhurt. I retired, loaded again and fired a second time, still he moved towards me. But when I fired the third time he turned back, and at the fourth shot he fell. This, my first success in ox-hunting, happened on the 23rd of August. While cutting it up, two officers caught sight of me and I heard one of them cheerfully exclaim: "Look there, Hans has got an ox!" After having stayed a while with me they said: "Better remain here, we are going on board to give information, and then the sailors will come to fetch it." When they had gone it grew evening, I spied with my glass, but nobody appeared. At last I repaired to the ship. I was asked: "Where are the men who went to fetch it,?" I replied: "I did not see anybody, I was tired of waiting for them and went off." I was informed that they were gone to bring me food. Not before I had been down to take my meal and came on deck again they returned.

The next five days I continued hunting, but in vain. On Sunday we attended divine service. The following day, August 30th, I travelled six English miles into the firth, crossed a hill, and, spying with the glass, discovered one ox. I descended towards the head of the firth and saw a herd of the same kind. Unfortunately my ammunition was insufficient, I had only nine charges left. However, I came up to them, shot six, and as then my ammunition was used up, I pelted one more [which was wounded?] with stones, but without killing it. I then returned. The day after I did not go out, but the next we set out to skin the oxen. We were accompanied by two officers who intended to go a-hunting in another direction. When I had climbed the hill I waited for my companions, and on their coming up we found the one I had pelted now dead. Its flesh began to turn on account of the entrails not having been taken out. Towards evening when they were going back, I said: "I will remain here until I have skinned them all." "But how wilt thou sleep?" "The ox-hides will afford me sufficient covering." When we had cooked and eaten they left me. In the evening when I had finished, I went to sleep. On awaking I saw three hares, I seized my gun and fired, but without hitting, whereupon I lay down again. After a while I rose, breakfasted and joined the ship.

On the 7th of September I got one ox, and only two hares on the 25th, and five in the beginning of October. When ice had formed, our Captain wanted to travel by sledge. We also went in search of the other ship in three sledges, one of which was drawn by dogs, and went back. On the third day we tried by land, but were obliged to give up on account of the deep snow. Some of our party also made a trip in another direction, but soon returned without having seen a live thing. I have noted my game during this season as follows:—October 12th, one hare; 13th, one; 17th, one; on the 16th, the sun was just to be seen [the last time]—18th, two hares; 19th, two; 22nd, one; 23rd, one; 26th, one; 27th, one.

In the month of November, moving about became difficult from the darkness. About this time I ceased to keep a record of the days. As I could not go out hunting more, I lent a hand at work. We built up walls of snow, making a large house for amusements [performances], and another large building we formed out of ice, intended for a smithy. The iceberg from which we fetched ice for drinking water was half an English mile off. I sometimes joined the men who had the charge of this, as I could not stand having nothing to do. I was not engaged for sailor's work, but only as hunter, sledge-driver and dog feeder. This is what I had promised, on leaving my home.

For three months I had no work at all. When daylight appeared and one could look for some game, I again tried for hares: the first time I got three, then sometimes two, and sometimes one. Already before the sun began rising again I was on my legs. I also did duty as the Captain's sledge-driver in surveying the country and climbing the hills, but when he remained at home, I went alone.

While the dark season still lasted I began to perceive that some of the crew were talking about me, and had wicked designs towards me. We also used to collect at nine o'clock in the morning, and stand upright in a row near the ship in military fashion. But I being a native was not accustomed to this. Two officers then proceeded to examine our faces, arms and feet [?] A little after nine o'clock the clergyman appeared to read prayers. This was repeated every day. Also in the evening they assembled to be inspected, but then without divine service. One evening I heard them talking thus: "When Hans is to be punished, who shall flog him?" The boatswain answered: "I." To be sure, as I am not very clever in English, and do not know whether I have thoroughly understood their meaning, I only have written this without any particular purpose

I also remember that in the beginning, when we took our meals, I was placed at the table of the first-class sailors [petty officers?], but afterwards I abstained [?] from their table. In this way I grew dejected, and the sadness of my mind was increased by my having no business on account of the terrible darkness. So when I took a walk near the ship I used to fall a-weeping, remembering my wife and little children, especially that little son of mine who was so tenderly attached to me, that I could not be without him even when I was travelling with the transport boat. However, I had one friend, a young man named Tage;[3] he sometimes took a walk with me, and when I made him know my sorrow, he consoled me. But at length my thoughts grew on me, and I took it into my head to go away to the wilds. If I should freeze to death it would be preferable to hearing this vile talk about me.

Once when heavy with grief I thus walked alone, I again heard them gossiping in their wicked manner. I then said to myself: "These people are all united as countrymen. I am the only one without any comrade of my nation, the only abandoned one"—and I ran away in the black night a distance of about 5 miles, when I stopped and meditated: "Our Captain likes me; perhaps he will send people in search of me; I will return, and if I am to be treated ill, the All-Merciful will pity my soul." I turned back but resolved to stop in the neighbourhood of the ship, as I knew our Captain who was

---

[3] Page, Captain Stephenson's steward.

my friend would search for me. I went ashore, dug a hole in the snow and lay down. I had just fallen asleep when I heard footsteps, and as they approached I went out. They asked me: "Hast thou slept?"—"Yes, I have slept." "Dost thou feel cold?"—"Only a little." The two officers said to me: "Our Captain was afraid thou wert lost, we have followed thy footprints, go home now." We went down upon the ice and met several men carrying torches. On coming on board the boatswain accosted me: "To-morrow morning sleep sufficiently; when we rouse thou needst not rise, only sleep in peace." The next morning I did not rise before I had slept well. At noon I got something to eat, and towards two o'clock I was summoned by the Master, who questioned me: "Why didst thou run away last night?" I made answer: "I heard them talk badly about me, and thought they reviled me." He rejoined: "Whenever thou hearst them speaking thus, tell me directly." I afterwards heard them speaking several times in the same way, but, nevertheless, did not mention it, because I supposed that, if I reported it, none of them would like me more.

When bright daylight had set in, the Captain and I used to travel about by sledge, to measure the height of the mountains. Once, on our way homeward, we saw a Tuluk coming. On drawing nearer, he shouted to us that a sledge had arrived from the other ship. When we returned, the officer next the Captain, Mister Bluman,[4] reported that Petersen had been frost-bitten, and that both his feet had been cut off. First they had departed, but turned back again on account of his being frost-bitten. All the others were well, they said. Coming on board, we met the four Tuluks. I made inquiry about my dear countryman on board their ship, and it gave me pleasure to hear that he was well. A few days later we set off for the south with three sledges, two of them drawn by men, while I, with the Commander, drove in a dog-sledge. We were five persons in one sledge. The two hand-drawn sledges were

---
[4] Lieut Beaumont, R.N.

intended for carrying provisions, to make deposits in the uninhabited tracts. The Captain and I returned, I have forgotten how many days after, leaving the others, who proceeded still further south. Having spent three days on board, we started to visit the other craft. We were five now also, and it took us several days to go thither. When I got to the ship, and saw my countryman, he appeared to me like a brother. He went off the next morning, and was absent sledging, I believe, for three days, and we then set off on our return journey. We fell in with a tent, a party with the boatswain as their leader, who had crossed to a part of our country [Greenland]. The other party, in charge of the officers, bad continued their trip. We stayed for awhile with them, and then went on. When we came to the rough ice we stopped, the Captain saying to me: "Look well after the road we have to go; if thou thinkest it is possible, we will proceed to-morrow." When we had eaten we went to sleep here. Early the next morning, while the Captain slept, I rose and went up the hills to con the ice. From the top I discovered that farther off the shore it was quite smooth. When I came down to the tent I found our cook at his work. The Master asked me: "How is our road?" I answered: "Farther outside it is excellent." After breakfast we started. When we had passed the hillocks we came to level ice, advanced quickly the whole day and went ashore in the evening, reaching our ship the following day.

Some time later we set out for the harbour where we wintered [Polaris Bay] when I was engaged with the Americans, on the coast of our country [Greenland], whereas now I was wintering on the west coast. We crossed the sound and arrived in three days. We found the house and put up our tent at its side. We took out the provisions [?] and examined them, the bread, the casks with smoked beef, one with molasses, also onions and several other kinds of eatables, whereupon we returned.

MY FOURTH VOYAGE TO THE NORTH, WHEN I WAS ENGAGED BY THE TULUKS

Afterwards we went off in three sledges. First, two drawn by men and carrying a boat; then we others followed three days after in a dog-sledge. At this time the sun began to give warmth, for which reason we slept by day and travelled in the night. We left in the evening, slept once and then reached the others who moved on foot drawing their sledges. We joined them and went in company to the opposite coast [Greenland] where we put up three tents on a cape, it was very pleasant indeed. Here we spent many days waiting for the party from the other ship, I had to drive with two officers, the second boatswain Fransmand Telle[5] and Mister Fulfut[6] and Kapine,[7] a young surgeon [?]. When at length they arrived, four sledges started on their return [?], whereas I with the two officers and the second boatswain set out for the south, to explore the interior of the firth.

At first we got along quickly and slept on the smooth ice. The next day we reached the glacier at the head of the firth. On the following we went out on foot, the two officers and I, along a steep cliff. In returning, the officers said to me: "Go to the tent and say: the officers want thee as cook; they are coming and will soon be here." I did as ordered, and when they came in, we ate and rested, smoking, whereupon off again to the glacier. During the night we walked over it [?]. In the morning when the sun gave warmth, we rested on the glacier. While the officers took a walk, I went off shooting, as I expected there might be a bear. I left my own gun and took along with me a breech-loader belonging to the ship. I went down upon the ice and came to a fissure. Here I shot at a big seal but missed it, when it emerged I missed it again. Next day I went off with my own gun and met with a large seal lying upon the ice. I crept towards it hiding myself behind my shooting

---

[5]Frank Chatel, Captain of the Forecastle, H.M.S. 'Discovery.'

[6]Lieut. Fulford, R.N.

[7]Dr. Coppinger.

curtain, and shot it, whereupon I went to fetch the sledge and dogs. The man who had remained came to assist me; we boiled some seal flesh, and had an excellent meal.

Our two officers had tried to walk over the glacier but found it very difficult, its surface being both slippery and terribly traversed by fissures. Before we started I went to fetch some of my seal flesh, whereupon we repaired to the edge of the glacier and rested there. The next day I brought the rest of the seal to the shore, and we crossed to the opposite side of the bay over an even ice field. Also from this side we found the glacier difficult of access, on account of its ruggedness. From hence I went with only one companion, Mister Fulfut, to look for the provisions we had deposited last year in a place thereabout. We found one bag with bread removed from the others; I think a bear must have tried to bite it. We returned to the rest and soon proceeded to an island in the mouth of the firth. I went hence with one man to fetch my seal flesh. On returning I got sight of an Utok [seal upon the ice]. I passed our tent, made for the seal and succeeded in shooting it, but left it to be taken away afterwards. The next day we examined the island.

From this spot we travelled to where I had wintered some years before, and where our house still stood [Polaris Bay]. We camped on the ice. On our arrival we found a sledging party without dogs, headed by an officer who last year had been with our other ship. They were in a pitiful state, suffering from scurvy, one of them having died, and only the officer and one man being able to walk properly. We made this place our temporary settlement, while I undertook to catch seals for them. The doctor ordered them to eat seal flesh to recover strength. At the same time we expected another party under the first lieutenant. After several weeks, as we began to grow anxious, I set off to look for them, accompanied by an officer, named Mister Rulsen,[8] and the doctor. We travelled all night,

---

[8] Lieut. Wyatt Ranson, R.N.

MY FOURTH VOYAGE TO THE NORTH, WHEN I WAS ENGAGED BY THE TULUKS

and when we approached our resting place, I left my companions, to look for the provisions placed there by me the other day. On my way back I fell in with a hatching ptarmigan. I seized it and likewise took the eggs, I believe there were six, and they were without young.

Next day we continued our searching for the missing party. I told my fellows I preferred driving overland, as there was too much snow on the ice. We took the strand and followed the shore-ice, but found it frequently inundated and almost impassable on account of the streams from the hills. My companions therefore walked over the land, while I continued along the shore-ice. They soon shouted to me to stop, as it was time to dine. Consequently we halted and took to cooking. When they came down to me, I said; "Look at that black point yonder, it appears to be a tent, and close by it I see something like a sledge." I grasped my spy glass and to our great joy made out the tent, the sledge and the two men.

When we had finished our meal, we left our tent and hastened to them. On drawing nearer, they came towards us, three men pulling a sledge. They stopped, and one of them advanced. We soon recognised him, Mister Bluman,[9] the officer next the Captain. He reported that they had four men sick, two of them on the sledge, two in the tent. Moreover, of the remaining three, one could scarcely walk, so there were hardly more than two to pull the sledge, and the leader looked very emaciated. We came up to them, and found their condition appalling. After having taken two on my sledge, I brought them to our tent, whereupon we fetched the others, with the tent and some provisions, my companions assisting by dragging. When we had rested, we removed to the place where we had our provisions [?]. I first had two [sick men?], and then returned to take the other two, just as the day before. Here we stayed awhile. I went out to shoot seals on the ice, but got none, on account of the deep snow mixed with water.

---
[9] Lieut. Beaumont, R.N.

We started, I carrying the luggage on my sledge, I was obliged to stop at times, to enable them to come up. I therefore proposed to fasten the other sledge to mine. When we had done so, we moved on quickly, both sledges being dragged by the dogs, and the hindmost, moreover, pushed forward by men. We travelled the whole night; when the sun began to grow hot, we rested. Next morning I started, with the doctor and two sick men, to reach our friends, whom we had left at the late wintering station of the Americans. We travelled the whole night under great difficulties, on account of the streams running down from the land and the inundated valleys. Having arrived, and given rest to the dogs, I went back again with two men; but before we started, one of the ailing men whom I had brought died. First we travelled over the ice, but as we were unable to proceed for water, we went ashore. I drove as much as possible over snow, although it was very soft and filled with water from below, whereas my companions preferred a snow-bare road.

When we reached the tent, the two officers came out to us, and when they were informed about the one who had died, they asked us not to mention this to their patients. When we had eaten we lay down in the resting-place of the officers. One of them wanted to use my dogs to fetch what they had deposited. When we awoke he had already returned, and was sleeping upon the sledge. We placed the two sick men and the tent on the sledge, and started with two sledges, both of them loaded [?]. When we were to go down upon the ice we left the other sledge and reached the house.

We now had three tents here a great part of the summer. I caught seven Natsek and three Ugsuk seals. Their flesh was a sort of medicine to the invalids. In June, hunting on the ice was hindered by its being covered with water. As soon as this began clearing off, I said to our officers that it would be better to cross the sound with the dogs before the ice broke up, as there was not sufficient room in the boat. They consented, and

MY FOURTH VOYAGE TO THE NORTH, WHEN I WAS ENGAGED BY THE TULUKS

I believe two days after they went off [?]. I followed to assist them, but returned to hunt for those who remained. Four days after their departure the ice broke up; I believe it was the day after they had reached the opposite shore.

Thereafter we waited many days for those who should come to fetch us. At lengthwe sighted people dragging a sledge and a boat. It was our Captain, who arrived with a number of sailors. After having stayed some days, he said to me: "To-morrow we will repair to the ship; go with us as our guide." Next morning we went off, leaving the sick, who had begun to walk about. When we were going, our Captain said: "Now, show us the road; go ahead of us, and we will follow." Thereupon we started, and crossed the open water in a boat. When we came to the heavy ice, I searched for the best road, accompanied by the Captain. He used to question me: "Which way are we to go?" I answered: "Look here; this will be better." It was lucky the Commander treated me as a comrade; I did not feel shy in speaking with him, as with other gentlemen. So we travelled partly over ice, partly by sea; I don't know how many days it took us to cross the sound. When we had reached the Westland, we proceeded by land to the ship. On approaching it we met with two men who were sent out to bring us something to eat—one of them the shoemaker, the other an engine-man. We rested and cooked, glad to see each other again, both of them belonging to my particular friends. After having finished our meal, we proceeded on board.

After reposing for two days, I went out to look for hares and wild geese. There were also many owls in this place, young as well as old. Of hares and geese I sometimes got as many as I could carry on my back [?]. But pretty far off the Tuluks had a tent on the top of a hill, to watch the approach of the other ship, which was now expected, and at the same time look for the men we had left on the opposite coast. Once it was said our other vessel was close by, only stopped by ice. This gave us much pleasure, the boat which was expected from beyond the sound now being our only care. Again they reported that our other ship had run ashore, but would be able to get off at high tide. Others believed it could not get off. To our great joy, it arrived. Now people were sent to fetch those on the opposite coast. They were turned back by ice, but then the others arrived of their own accord, and now at length we were ready to start on our return voyage, though the ice was still very bad.

The Captain of the other vessel kept a good lookout for the road; he went up the hills incessantly without being tired. At length we started, although it appeared as if there was no thoroughfare in the direction we were to go. However, farther off there was more open water, and this we reached. During one day and night we traversed a tolerably ice-free sea, but then again had to force our way close along the shore, going on whenever the pack retired a little. The Captain of our other ship was beyond all praise, one might think he neither slept nor ate. Sitting in his look-out in the mast he sometimes took his meal there. On account of his extraordinary skill in ice-navigation he was our leader. During this passage I caught two seals, one Natsek and one Ugsuk. While we thus struggled with drift-ice, new ice began forming, thick enough to be walked over, but at length we came to open sea.

During the night they wakened me and gave me a letter; I directly recognised it to be from Upernivik, from my homestead. When I had read it, I learned that my wife and children were well, and now I felt consoled. We proceeded against a southerly wind and searched for whaling vessels on their fishing grounds, but did not see any. We sighted land about Tasiusak and tried to touch at Natsilivik, where lived natives, but were obliged to turn seawards, and to my disappointment passed Upernivik without approaching the coast on account of the heavy gale. When the wind abated we landed at Kekertarsuak [Disko]. Here I was allowed to remain, and I felt consoled to know that I could stay with the Inspector, as he was very friendly towards me. He desired me to write what I had seen, and though unskilled in composition, I have tried to give this account of my voyages, while engaged thrice with the Americans and once with the Tuluks. Four times in all I travelled to the North.

And now I bid farewell to all who have read my little tale. I minded my business, sometimes under hardships, sometimes happy. May all who read this live happily in the name of the Lord!

Written in the year 1877.

* * * * * * * * *

Note.—When I wrote the present account, I knew the 'Polaris' Expedition only from its being occasionally mentioned in various journals and books, especially Commander Markham's *A Whaling Cruise*. After its being delivered to the press, I was favoured by receiving from the Naval Observatory in Washington a copy of the official work on the said expedition, which, of course, I immediately studied with the greatest interest. It may be understood how I was gratified by seeing how Captain Tyson and his fellow-sufferers fully acknowledge the assistance rendered to them by Joe and Hans, without whose skill in catching seals they would never have escaped perishing from cold and starvation during their marvellous passage across the ocean upon a piece of ice. The disagreements found here and there between the two accounts I find are comparatively trifling, and easily explained by so many details being given by Hans merely from his memory. A curious one is that "Mister Tarsta" seems, to refer partly to Chester, partly to Tyson. Whether he may have taken the names to be identical, or confounded the persons in his memory, I am quite unable to make out.

H. R.

Printed in Great Britain
by Amazon